SECURITY AND SAFETY IN CRUISE TOURISM

SECURITY AND SAFETY IN CRUISE TOURISM

Agatha Bierce

ABD PUBLISHERS
Jaipur ■ New Delhi

ABD PUBLISHERS

Regd. Off. "Bony Residency", Gate No. 2
Opp. Tilak Public School, Vishweswariya Nagar,
Gopalpura Road, Jaipur - 302018, India
Ph: 0141 - 2761280, Telefax: 2761381
Email: abdpublishersjpr@gmail.com

Delhi Office: 102, 1st Floor, Satyam House,
4327/3, Ansari Road, Darya Ganj,
New Delhi - 110002, India
Ph: 011-45652440

visit us: www.oxfordbookcompany.com

Distributed By
KEYA INTERNATIONAL PRESS LLP
E-mail: keyainternationalpressllp@gmail.com

Published in India, 2026

ISBN: 9788183768573

Typeset by Shivangi Computer, Jaipur

Printed and bound in India by Thomson Press Ltd.,
New Delhi

Contents

Preface

With more than 13 million passengers cruising each year, the cruise line industry goes to great lengths to maintain the safety and security of our passengers enjoying their cruise vacations. We are proud of our safety record and our long-standing partnership with law enforcement agencies, including the FBI and U.S. Coast Guard. Cruise lines meet and often exceed numerous international and federal codes and regulations designed to maximize the safety of passengers aboard ships. In July 2010, a new law that the cruise industry supports went into effect that contains some new provisions and clarifies other existing CLIA policies that help ensure cruising remains safe. In addition to announced and unannounced safety inspections, the U.S. Coast Guard annually conducts a Control Verification Certificate examination for every cruise ship that ports in the U.S. for compliance with both federal and international regulations. The cruise industry has a Security Committee, made up of security officers from each cruise line, which meets every 60 days with law enforcement and intelligence agencies to discuss relevant issues such as port and ship security and intelligence assessments. These agencies include the U.S. Coast Guard, FBI, Department of Transportation, Customs and Border Protection, Office of Naval Intelligence, and Department of Homeland Security, among others.

The Cruise Lines International Association (CLIA) and its members go to great lengths to ensure the security of its passengers, crew, ships and maintain compliance with all regulations. Cruise ships are safe and secure environments with trained, 24-hour security personnel. Over the past two decades, an estimated 90 million passengers have enjoyed a safe cruise vacation. In the rare instance of an allegation of crime, member lines follow comprehensive policies and procedures. These policies are working. In fact, in 2007 the U.S. Coast Guard testified before the U.S. Congress that these policies are working well and that there is "no data to suggest that crime on cruise ships is more prevalent or sever than in any other vacation venue." (Coast Guard testimony) Serious crime aboard cruise ships is rare, however, even one incident is one too many. The industry has a zero tolerance policy when it comes to crime. As required by law (*33 CFR 120.100, 120.220* and recent legislation signed by President Obama), CLIA member lines report allegations of crime occurring in U.S. waters or on a vessel operating to and form a U.S. port to appropriate law enforcement agencies, such as the FBI. We also worked with the FBI and U.S. Coast Guard to agree upon a mechanism we're using today to report all alleged crimes involving U.S. citizens anywhere while at sea regardless of ship registration.

This book contains advanced information about this subject. This book will be a boost for the learners and an essential subject manual.

—***Agatha Bierce***

1

Personal Safety and Security on Board

Arriving by Cruise Ship-tips on Having Fun

Discover the Truth about Personal Safety in Cartagena

Each year, hundreds of thousands of tourists arrive at our modern Port for a brief visit to Cartagena. Cartagena de Indias was the most important city on the main sea route, then known as: "The Spanish Main". The city held vaults full of gold, silver, emeralds and other treasures looted from the indigenous peoples of the continent, and Central America. Annually, when good sailing months permitted, the hoards were shipped to the treasury of the King of Spain. Because of her stored wealth, Cartagena was high on the "hit list" for the pirates.

Entering the Harbour, is in itself, an interesting part of your visit, and traces the path of the early galleons. Your first indication that you are approaching Cartagena occurs miles from the city, as your ship majestically glides right between two ancient Spanish Forts near Bocachica (Small Mouth), and then mid bay between the Forts of: Castillo Grande and Manzanillo. As pirate ships appeared on the horizon, a under water chain was raised from the sea bed between the forts, tightened and held fast, with a large capstan. From this advantage the ship was pummeled with cannon shot from the two opposing forts. As you continue your approach to the city you will see additional fortifications including the mighty Fort San Felipe de Barajas, which arguably is the largest and most impressive fortification in the new world..

The Port of Cartagena is one of the most modern in South America and one of the busiest in Colombia. The Port cooperates fully with Interpol (of London, England) and the DEA (Drug Enforcement Agency) of the United States. There are video cameras everywhere, as well as a large contingent of well armed and trained police, plus drug sniffing dogs. Please do not be alarmed. The "troops" are friendly to tourists, and they will help you in any way that they can. A primary reason for this heavy security results from the insatiable demand for drugs consumed in the United States and other prosperous countries of the world.

The first law of Economics is: supply will meet demand. First, there is demand. The U.S. problems relating to drugs is shared by the people of Colombia, where more than 40,000 have been killed in the past 35 years. Within this same period, a much higher number of drug related deaths, have occurred within the United States. Both countries share a "bad name" when it comes to drug violence and killings.

Your visit to Cartagena is very much appreciated, and a big help to the local economy. The vast majority of locals truly love and appreciate gringos, and others foreigners who visit their city. There are more than 80,000 Colombians residing within the United States. It is a sad note, but it can be safely said: the majority of youth in Colombia, would love to: "live and work in the states". For many, it is their daily dream!

Some of your shipmates will not get off the ship in Cartagena. Others, who are equally un informed, will venture off the ship, but remain in the port, so they can brag to their friends: "I was in Colombia". The usual response will be: "you were where?...are you crazy?"

Cartagena is not representative of Colombia, and there is little to worry about when it comes to your personal safety. The city, and in particular, the old walled portion called El Centro, is clean, safe, fun and romantic. Those who are not as informed as you, will be the big losers if they miss this golden opportunity to visit one of the greatest cities in South America.

While on your Cartagena adventure, you will see poverty-up close. The U.S. State Department for years, has issued a: Travel

Warning to U.S. Citizens against visiting Colombia. If you seek a more balanced and accurate report, visit the Canada Advisory. Canada has a consul in Cartagena. They know first hand about personal safety. The Travel Warning issued by the U. S. State Department is seriously hurting the economy and people of Cartagena! If you are a U.S. citizen, you should feel good knowing that your visit will help our economy. You are welcome, and the people of Cartagena thank you.

Note: This text is written by a 62 year old U.S. citizen who lives and works in Cartagena. "Even with a so-so command of the Spanish language, I don't feel that I have ever been in personal danger in Cartagena. My work keeps me out at all hours, visiting clients, many times while I am alone. Of course, I am cautious as to where I go, the way I dress and act, but this is also true when I visit my home state of Florida." Use a little common sense, and you will have no problems.

Cartagena, is the most popular tourist city in Colombia. Our poverty is demonstrated by the high number of street hustlers. Foreigners stand out. You can be spotted a block away. When you are in a crowd, such as on a shore tour, the word passes quickly, and your group will be besieged like bees heading for the flowers. Hustlers are poor, and tourists represent the rich honey pot-their next meal.

You will be safe, and generally, you will not even be touched, including a friendly hand on your shoulder. If you give the hustlers, beggars, or children as much as a single coin, you will be hurting them. They have no pride. You cannot humiliate them, for they are at rock bottom. If you have no pride, hustling and begging is easy. If you give them anything, they will be encouraged to continue their ways, looking to the tourists for support. Their presence in the streets of Cartagena, is seriously hurting the tourists image of our great world heritage city, causing great and lasting economic damage.

(Note: Colombia spent more than a million dollars to "clean up the streets" when President Clinton visited Cartagena. However, they have no funds for a continuing effort, or solution, to help the poor. Meanwhile, the U.S. government sprays poison upon the poor farmers of south Colombia, rather than adopting a sustained commitment to help with needed education, and improved economic opportunity.)

These street hustlers and beggars have other means of honourable support, such as: "work". As an example, look at the street peddlers of Cartagena. They sell products and offer services to both the people of Cartagena, and the tourists. They work long and hard. Their honest work provides a poor, but adequate income. The cost of inventory for a street peddler is very low, usually less than $10.00. If you give the hustlers or beggars money, many will quickly spend it on: lotto tickets, alcohol, cigarettes, slot machines, drugs, etc. Many of the barefoot beggar kids buy glue to sniff-which permanently damages their brain and nervous system. You can easily spot the glue kids. They wear long sleeve shirts which hide their stash of glue. When they raise their wrist to their nose, they are getting a "fix"-inhaling the deadly fumes. Give the kids shoes, and they will sell them, perhaps to buy again, from the above list? The reason is simple: if they wear shoes they will not look poor enough, and their begging income will quickly drop. Some beggar women rent children to pose as their own. They carry the child in their arms, begging. They point out the child's deformities, seeking your sympathy-and then your money! Note: The following is sad but true: Many of these "rental children" have been intentionally maimed, such as an eye gouged out, an arm cut off etc. The reason: it raises the rental value of the child. The more pathetic and sickly the child, the more the "loving and caring" tourists will give. Tourist money has been the cause of many maimed children-the world over, it is not just a Colombia problem.

When tourists, become more informed and involved in supporting lasting solutions, this may stop. We have hope that the Internet will be a primary tool in this educational effort. In the mean time, simply giving money, only continues this deplorable situation.

Don't be upset about what you have just read. The people of Cartagena thank you for your concern over their economic problems, but above all, they want you to have a good time-for that is the reason for your vacation! So back to your shore visit. You can take the ships shore tours, or you can go it alone. Either way to visit Cartagena is fine. When you take the bus tours, you will be escorted to places that may not interest you. There will be many sites where you want to stay longer, (or shorter), but you have no choice, you go where the bus goes. Additionally, you will not be able to hear all that the tour guide has to say. And, you may have questions that will remain un-answered? However, these tours can be a loads of fun, and many enjoy them. There is a touring alternative. A private guided tour!

By far, the best and most positive reports appearing on the Internet Newsgroups, come from tourists who have taken a taxi tour of Cartagena. Taxis will be waiting at the dock.

Note: Cartagena taxis are modern, compact, 4 door models, some with air at an additional cost of 10%. The streets of the old city are very picturesque, narrow, and therefore, just right for compacts. You will probably want to spend at least 75% of your tour in the old walled city? Costs per tourist will be minimum if there are four persons in your private group. Try this tip. As soon as you hit the dock, each of you head for a different taxi and quickly talk (mostly "listen") to the drivers. Test their English, or your desired language.

You are lucky to have visited and studied this web site, for here you can learn exactly what it is that you want to see, where you want to shop, or stop for drinks or a great meal. Print out the first page of our web site Walking Tour, and check off the things you want to see. Show this list to the driver. Don't forget to allow enough time for shopping. And if you or your group is interested in buying emeralds or gold, be sure to check out our extensive site: Buying Emeralds and Gold in Cartagena-Tips to Save you both Time and Money. Taxi rates are uniform, and they are printed on a plastic card carried in the glove box of all Cartagena taxis. Tip: Ask to see the card, and then ask the price per hour, and total based on the minimum number of hours, for your planned tour.

There is so much to do and see in Cartagena, and your time is very limited. This web site is dedicated to helping you make the best of your time and your money, so you will fully enjoy your Cartagena experience. A little planning on your part will go a long way. When you return home, you may want to tell others on the cruise news group of your Cartagena experience?

Cartagena is a great romantic and historic city, certainly worth at least a week of your time. Who knows, you may want to stretch your pension and live the good life, here in Cartagena on the Caribbean!

Safety and Security

With more than 13 million passengers cruising each year, the cruise line industry goes to great lengths to maintain the safety and security of our passengers enjoying their cruise vacations. We are proud of our safety record and our long-standing partnership with law enforcement agencies, including the FBI and U.S. Coast Guard.

Cruise lines meet and often exceed numerous international and federal codes and regulations designed to maximize the safety of passengers aboard ships. In July 2010, a new law that the cruise industry supports went into effect that contains some new provisions and clarifies other existing CLIA policies that help ensure cruising remains safe. In addition to announced and unannounced safety inspections, the U.S. Coast Guard annually conducts a Control Verification Certificate examination for every cruise ship that ports in the U.S. for compliance with both federal and international regulations.

The cruise industry has a Security Committee, made up of security officers from each cruise line, which meets every 60 days with law enforcement and intelligence agencies to discuss relevant issues such as port and ship security and intelligence assessments. These agencies include the U.S. Coast Guard, FBI, Department of Transportation, Customs and Border Protection, Office of Naval Intelligence, and Department of Homeland Security, among others.

Regulations and Compliance

There are numerous international and federal codes and regulations with which the cruise line industry must abide. It is important to note that all ships visiting ports, regardless of where they are flagged, must comply with all applicable federal regulations. In the U.S. for example, according to the Coast Guard website, they have "the responsibility for ensuring that these ships meet all international conventions and domestic requirements for safety, security and environmental protection." The Cruise Lines

International Association (CLIA) serves as a non-governmental consultative organization to the International Maritime Organization (IMO), an agency of the United Nations. The IMO mandates global regulations for the safety and operation of cruise ships. The most important of these is the International Convention for the Safety of Life at Sea (SOLAS), which includes comprehensive information on safety equipment and procedures. Following the September 11, 2001 attacks, the International Ship and Port Facility Security Code (ISPS), a broad set of security requirements for ships and port facilities, was added as an amendment to SOLAS. Compliance with SOLAS standards and other internationally recognized conventions is monitored by both the ship's flag and port state representatives, which is the U.S. Coast Guard in the U.S. In July 2010, President Obama signed the Cruise Vessel Security and Safety Act into law. The cruise industry fully supports the law and in fact worked with Members of Congress as the bill was being drafted. It contains some new provisions and clarifies other existing CLIA policies, such as how the cruise lines report allegations of crime to the U.S. government. While many ships in the CLIA fleet may already feature 42-inch railings, peepholes in all cabins, onboard video surveillance systems, and employ medical personnel who meet guidelines established by the American College of Emergency Physicians, the new law will make these and other new provisions consistent across the fleet.

In addition to announced and unannounced safety inspections, the U.S. Coast Guard annually conducts a Control Verification Certificate examination (with quarterly reinspections) for every cruise ship that ports in the U.S. for compliance with both federal and international regulations. This effective oversight system means, for example, if the U.S. Coast Guard finds a cruise ship to be in violation of any required regulation or considers it unsafe in any way, the local Coast Guard Captain of the Port has the authority and responsibility to prevent passenger boarding or departure from a U.S. port with passengers onboard until those deficiencies are corrected.

Security

The Cruise Lines International Association (CLIA) and its members go to great lengths to ensure the security of its passengers, crew, ships and maintain compliance with all regulations. Cruise ships are safe and secure environments with trained, 24-hour security personnel. Over the past two decades, an estimated 90 million passengers have enjoyed a safe cruise vacation. In the rare instance of an allegation of crime, member lines follow comprehensive policies and procedures. These policies are working. In fact, in 2007 the U.S. Coast Guard testified before the U.S. Congress that these policies are working well and that there is "no data to suggest that crime on cruise ships is more prevalent or sever than in any other vacation venue." (Coast Guard testimony)

Serious crime aboard cruise ships is rare, however, even one incident is one too many. The industry has a zero tolerance policy when it comes to crime. As required by law (*33 CFR 120.100, 120.220* and recent legislation signed by President Obama), CLIA member lines report allegations of crime occurring in U.S. waters or on a vessel operating to and form a U.S. port to appropriate law enforcement agencies, such as the FBI. We also worked with the FBI and U.S. Coast Guard to agree upon a mechanism we're using today to report all alleged crimes involving U.S. citizens anywhere while at sea regardless of ship registration.

When an alleged crime is reported, CLIA member lines' trained security personnel handle such situations with professionalism and follow established procedures that include detaining suspects and, if appropriate, securing a crime scene. Our security personnel have all received training on crime scene preservation from FBI Agents and a FBI-training DVD developed specifically for the cruise industry. If an alleged crime is sexual in nature, medical professionals on board have sexual assault evidence collection kits and are trained in their proper use. We are also working with the Sexual Assault Resource Service to expand current training on sexual assault evidence collection. All policies and procedures are well-established, but we also constantly review and update them, keeping our guests' safety and security top of mind.

Since cruise ships operate in a controlled environment, access can be strictly enforced. All passengers, crew, carry-on baggage and luggage are inspected prior to boarding. CLIA member lines maintain an official manifest for each ship, which includes everyone on board a cruise ship, all officers, staff and crew, as well as all guests. Both passengers and crew may embark or disembark only after passing through a security checkpoint.

All crew undergo rigorous pre-employment background screening. In addition, foreign nationals who work on ships that sail into U.S. ports are required to meet the legal obligations of a seamen's visa, as set forth by the U.S. State Department

CLIA member lines work in close cooperation with international and federal agencies regarding shipboard safety issues, including piracy. CLIA members maintain an extensive network of intelligence gathering among government and private sources. Based upon this monitoring, the cruise line industry is constantly assessing this information and any potential risks for member lines. CLIA members have thorough security protocols in place and follow international guidelines concerning regular contact with military forces to ensure ship and passenger safety.

Ship Safety

Cruise ships are designed and operated in compliance with international and federal regulations specifically designed to maximize the safety of passengers and crew. All cruise ships visiting ports, no matter where they are flagged, must comply with applicable federal regulations.

The Cruise Lines International Association (CLIA) has a unique partnership with the U.S. Coast Guard through their Cruise Ship Centre of Expertise in Miami. The U.S. Coast Guard has "the responsibility for ensuring that these ships meet all international conventions and domestic requirements for safety, security and environmental protection." The U.S. Coast Guard conducts announced and unannounced safety inspections for every cruise ship that embarks passengers in U.S. ports.

U.S. Coast Guard inspections include conducting plan reviews of each cruise ship before construction is even started, inspecting the ship at the ship yard during construction, conducting a comprehensive initial Control Verification Examination upon delivery, and annually conducting a Certificate of Compliance examination (with quarterly reinspections) for compliance with both federal and international regulations. This oversight system means, for example, if the U.S. Coast Guard finds a cruise ship to be in violation of any required regulation or considers it unsafe in any way, the local Coast Guard Captain of the Port has the authority and responsibility to prevent passenger boarding or departure from a U.S. port with passengers onboard until those deficiencies are corrected.

Inspections focus on structural fire safety, proper functioning of all safety systems and equipment including fire fighting systems, lifesaving equipment and other safety systems such as the lifeboats, life rafts and lifejackets in addition to crew training and competence. Additionally, the U.S. Coast Guard observes fire and abandon ship drills conducted by the ships' crew, and requires satisfactory operational tests of key equipment such as steering systems, fire pumps and bilge pumps.

All cruise ships have lifeboats, life rafts and life preservers for every person on board with additional capacity in accordance with international regulation. SOLAS requires that lifeboats be capable of being loaded, launched and maneuvered away from the ship within 30 minutes of the Master's signal to abandon ship.

The average CLIA ship, of approximately 97,000 gross tons carrying approximately 2700 passengers and 800 crew, typically has five firefighting teams whose main members have advanced shipboard firefighting training, 4,000 smoke detectors, 500 fire extinguishers, 16 miles of sprinkler piping, 5,000 sprinkler heads and 6 miles of fire hose. Sprinklers and smoke detectors, as required by international regulation, are located throughout the ship.

Industry Initiatives

There is no higher priority to CLIA member lines than the safety and security of passengers and crew. In coordination with the International Maritime Organization, U.S. Coast Guard, and flag and port states, we've undertaken several initiatives to enhance the numerous regulations and codes to which we adhere.

The industry has a special relationship with the U.S. Coast Guard through their Cruise Ship National Centre of Expertise. They have "the responsibility for ensuring that these ships meet all international conventions and domestic requirements for safety, security and environmental protection."

Our security committee, made up of security officers from each cruise line, meet every 60 days with law enforcement and intelligence agencies to discuss relevant issues such as port and ship security and intelligence assessments.

Although serious crime aboard cruise ships is rare, we have worked particularly closely with authorities to ensure we have properly trained security personnel who know how to respond appropriately to any allegations of crime. We instituted a zero tolerance policy when it comes to crime, and in addition to our mandatory reporting requirements, we also have an agreement with the FBI and U.S. Coast Guard to report all alleged crimes involving U.S. citizens anywhere while at sea regardless of ship registration. Lastly, we worked with the FBI to provide our security personnel with a comprehensive crime scene preservation training DVD, developed specifically for the cruise industry.

Safety Risks the Cruise Industry Doesn't Want You to Know About

Summer is just around the corner, and, despite an array of travel options, many people choose to spend their holiday on the water. Each year, over 10.2 million U.S. citizens take trips on large cruise ships, and, for the most part, voyages are safe and fun. However, just like any other mode of transport, there are inherent risks, many of which the cruise industry would prefer not to be highlighted by the media. Annually, an average of 150 people fall overboard, and several others suffer personal injury, outbreaks of disease, and even death. While fortunately no passengers were seriously injured during a mishap on a Carnival cruise liner that recently departed from California, the incident illustrates some of the risks, explains a lawyer in the state.

Last November, an engine fire on the 113,000-ton Carnival Splendour took out the ship's power. Its 4,500 passengers were stranded for several days off the western coast of Mexico without air conditioning, hot food, flushing toilets, and telephones until tugboats arrived to pull the ship to San Diego. Although no one was seriously hurt, at least three people were treated for panic attacks. Had the Splendour been further out to sea, it may have taken much longer to tow it to port, and conditions that were inconvenient for a few days may have become safety hazards after a few weeks as supplies were exhausted.

In another incident involving Carnival Cruise Lines, a ship called the Ecstasy returned to Galveston, Texas with 60 injured passengers after it abruptly listed during a manoeuvre to avoid a partially submerged buoy while sailing near Mexico's Yucatan Peninsula. Fortunately, there were no life-threatening injuries reported.

In addition to accidents onboard the ship, the virus known as "Norovirus" has been known to spread throughout vessels at sea quickly, causing vacation plans to be ruined. Avoiding others who appear sick, washing your hands often, and staying healthy can reduce the risk of contracting the virus. Although some people recover after a few days, in serious cases hospitalization is required to avoid further medical issues.

The Centres for Disease Control and Prevention, CDC, has created the Vessel Sanitation Program, in which the inspection reports for all cruise ships are published. If you plan to book a cruise, take a few minutes to check out the ship's score. The worst scores are in the 60s, but most new ships score in the 90s. About seven or eight ships actually have a perfect 100, indicating that the food is kept at appropriate temperatures and safety features are well maintained, according to a lawyer. While accidents like the ones that forced the Splendour to return to California early and caused personal injury to the passengers of the Ecstasy are unpredictable, researching a ship's sanitation score will at least reduce the risk of spending your next vacation sick.

How Safe is a Cruise?

The debut of two brand-new new cruise ships — Cunard's Queen Elizabeth earlier this fall and Royal Caribbean's massive 16-deck Allure of the Seas in December — coincides with the beginning of "wave period," a time of year when most people book their cruise vacations.

But talk of cruises inevitably raises the subject of cruise safety. A few weeks ago, a 79-year-old British man disappeared from a cruise ship in the English Channel. He's only the latest in a list of passengers who either vanished or fell overboard.

The cruise industry contends a trip on the high seas is safer than a drive to the airport and a stay at a hotel. But just how safe is it? There is little evidence that people are not booking a floating vacation because of safety concerns. But this summer, President Obama signed the Cruise Vessel Security and Safety Act, which requires cruise lines to report all deaths, missing persons, theft, sexual harassment and assaults. However, the Department of Homeland Security hasn't yet published the data and did not offer a timeline for its release.

Until that happens, experts say, safety can depend on your behaviour.

Stay in your cabin and order room service for the duration of your cruise, and short of your ship sinking, you'll probably be just fine.

But stay up all night drinking, do your "I'm-the-king-of-the-world" impression from the bow of the ship, and take the unauthorized shore excursion to the slums — well, maybe not.

That's not to say the cruise line isn't responsible for your well-being. It's just that there's only so much it can do to keep you out of harm's way while you're at sea.

Sexual Assaults

Cruise industry representatives claim sexual assaults are practically unheard of at sea — according to the industry, you have a less than 0.1 percent chance of becoming a victim onboard. However, as cruise expert Ross A. Klein testified before a Senate hearing in 2008, the rate of sexual assault on cruise ships is almost twice the U.S. rate of forcible rape — about 56.9 per 100,000.

"It is common knowledge in our field that cruise ships are flat-out dangerous for women," said Amelia Stinson-Wesley, the former executive director of a battered women's shelter and rape crisis centre. "Not as dangerous as countries at war where rape is used as a weapon of war, but more dangerous than even a college campus and certainly more dangerous than the general population of most cities in the United States."

Passengers can protect themselves by taking some common-sense precautions, including keeping your cabin door locked, not opening your door for strangers and checking in with a travel companion, according to the Rape, Abuse & Incest National Network.

Stomach Ailments

Gastrointestinal outbreaks spread easily from person to person and can occur aboard cruise ships or in hotels, nursing homes or day care centres.

But there's a reason the Norwalk virus is sometimes also called cruise-ship sickness: because it happens on cruise ships. Here are the ones reported to the Centres for Disease Control and Prevention (CDC). (These are just the ships where 3 percent or more of passengers or crew reported symptoms of diarrheal disease to the ships medical staff during the voyage.)

More people are at risk for gastrointestinal illnesses while traveling than ever, according to the CDC. "Cruise ship passengers and crews are also at risk for new patterns of old diseases or newly recognized causes of diseases," the agency warns. You can prevent an infection by frequently washing your hands, carefully washing fruits and vegetables before eating them and cleaning and disinfecting contaminated surfaces, according to the CDC.

Sick at Sea

Cruise experts say today's ships are so big, you won't even know you're at sea. Mention the word "seasickness" and those same insiders reassure you with talk of stabilizers and almost-always calm waters.

But waves still happen. And seasickness happens. Ann Azevedo took an Alaska cruise during the spring and on the way back, she encountered some rough weather. "The decks were shut down," she recalled. "The weather information on the TV stopped updating. We tried to go to bed, but there was no way we could sleep — we were near the bow, and every fourth wave or so would hit with a shattering boom. They were breaking across our balcony — and we were on the ninth deck." Rough weather can normally be avoided by steering clear of cheap, off-season cruises (the Caribbean during hurricane season or Alaska during the late spring or early fall).

Tightening Cruise Ships' Security: State of Access Control Solutions Onboard Passenger Ships

If you wanted to describe the security challenges of a cruise liner, you could think of a hotel inside an airport. When it comes to the individual cabins, a cruise ship requires the same kind of security as a hotel: secure doors to the rooms, records of who has been going in and out, perhaps surveillance cameras in the corridors to check on lurkers.

But while a hotel with its lobby and restaurants is usually open to the public, nobody should be able to get on to a ship if they are not entitled to be there. So boarding procedures have to be very similar to those at airports, with boarding passes, baggage controls and body checks.

Awareness of the vulnerability of cruise ships, says Geoffrey H. Greaves, CEO of consultants International Maritime Security in London, is not new. The Achille Lauro hijacking in 1985 led the International Maritime Organization (IMO) to issue guidelines for security on cruise ships. *"After that, ships, especially those going to America, had to have minimum standards,"* says Greaves.

But the urgency grew after 9/11, when people began to imagine that a big cruise liner could be the target of a terrorist attack, leading the US to issue its International Ship and Port Facility Security (ISPS) code. The code requires every ship over 500 tonnes to meet certain standards and ships are not allowed into the United States unless their last ports of call conformed to ISPS code.

New Security Rules for Cruise Lines

Greaves says the new rules only led to a slight change in procedures for cruise lines that had already been following the IMO rules. But he says modern ships are designed to include security features. *"The fibre optics for the cameras will be provided, and the wiring for the archway metal detector at the gangway will be installed,"* he says. *"I visited a ship under construction recently where I immediately saw that the gangway was too narrow. Nowadays a gangway has to be designed so that it's wide enough for all the security equipment."*

With passengers and crew embarking and disembarking at every port, the same procedures will be followed each time. That means maybe 4,000 passengers and crew having to go past a baggage x-ray, an archway metal detector, body checks, and a swipe card reader with a crew member sitting by a computer watching the picture of the card holder come up. The picture will be taken when the card is issued.

Access controllers could also use fingerprints as an additional check that the person who uses the card is the passenger, but nobody is using biometric fingerprint readers yet, says Preben F. Poulsen, Vice President of VingCard Marine, which makes ship security systems: *"There is the problem that not everyone has good fingerprints for reading, but we offer the option. Probably it will only come into use when a disaster happens."*

Analysing Security Risks for Passenger Ships

That a disaster hasn't yet happened surprises analysts. The number of passengers carried on the Cruise Lines International Association's ships rose from 7.2 to 11.2 million between 2000 and 2005. In 2002 the tanker Limburg was attacked in Yemen and destroyed, and there have been a couple of attacks on ferries in Indonesia, but so far cruise ships have remained untouched. Greaves says it's not because security is so good. *"I could get you on board a ship to carry out a spectacular attack,"* he says. He notes it's difficult to convince cruise line operators to increase their security as long as everything is going well.

But although one might think that passengers would want to imagine they were in a dream world where security was not an issue, in fact, they demand the safety of proper controls. *"I was standing behind a couple on a gangway recently,"* says Greaves, *"and there was nobody there as they went through the metal detector. The woman said to the man, 'Hey, John, I hope it's switched on."*

Stowaways are still fairly common, but their numbers have gone down. *"Someone should be at every open entrance to a ship,"* says Greaves. *"But sometimes the crew member goes round the corner to have a cigarette."* Access controls for crew have to be particularly stringent. Two stowaways got on one of his clients' ships recently pretending to be security officers. *"But the crew were able to use CCTV pictures to identify them and remove them at the next port of call."*

The security issues on board a ship are usually much more mundane than the risk of terrorist attacks. *"We have all kinds of crime, from theft to rape,"* says Greaves. *"In the early nineties, there used to be cameras only at the entrance. Now they are likely to be all over the vessel to monitor both passengers and crew."* There might be passengers who claim to have slipped in order to gain an upgrade, or accusations of sexual molestation by a crew member.

New Developments in Access Control on the Horizon

Developments in the cruise ship industry are often similar to those on land. For example, some security companies offer online locks for all security areas, or the cabin key-card can be used to make purchases on the ship. *"In fact, it was convenience which was driving things originally, but then came 9/11,"* says Preben F. Poulsen. But there are some marine specialties, he adds: *"For example, we always include a mechanical overriding cylinder in the locks in case of evacuation."*

Peace of Mind-Travel Insurance

I know-you don't need travel insurance-right? Okay, if you say so, but please humour me and read on...

You never know what might put your vacation investment at risk. The last thing you want to think about is what would happen if you became ill or injured while away from home.

Most travellers make sure they have some form of medical insurance before they leave home but what about all the dollars you have invested in your vacation?

No matter how carefully you plan your trips, illness, accidents and family emergencies can occur unexpectedly and quickly disrupt travel plans.

While most travellers never intend to cancel their trip or return home early, you still need to protect the investment you have made in your vacation – from airline tickets to hotels and beyond.

A Travel Insurance story...

"A traveller developed a sinus infection while on a cruise with her husband and when she got off the cruise, the infection was full-blown,"!

"They visited a doctor who told them there was no way she could fly since the pressure would be too much and they ended up staying four nights longer than planned."

Luckily, the couple had purchased cancellation and interruption insurance, which covered their one-way tickets home to Victoria, Canada and, since they had planned to visit friends for a few nights after their cruise, the unused portion of their vacation was also covered.

The total cost of their claim was just over $2,600 and paid entirely by the Insurance company. "Someone from the Travel Insurance company phoned the travellers, went through their claim with them and rebooked their tickets back to Victoria,"

"An Insurance representative even phoned them when they arrived back in Victoria. They were really impressed."

Fortunately, travel insurance can protect you from unforeseeable losses stemming from an array of circumstances. Some travel insurance even allows you to cancel your vacation for any reason at all. Many travel insurance companies offer cancellation and interruption insurance to protect against penalties customers would have to pay if they had to cancel their trip prior to departure, due to an insurable risk.

This type of insurance also provides coverage in the event travellers have to interrupt their trip and return early or stay longer, due to a covered situation.

Some examples of commonly covered risks include:

- The emergency medical condition or death of yourself or an immediate family member
- The emergency medical condition or death of your travelling companion
- Complications due to pregnancy
- A travel advisory issued by your home government after you purchased your trip, advising citizens not to travel within or to your destination country
- The delay of a scheduled carrier.

Which kind of Travel Insurance Do I Buy?

Buying travel insurance requires that you consider your needs carefully.

Check with your credit card company to see what insurance may be automatically provided by them if you use their card to purchase your vacation.

Some credit card companies only cover a portion of your risk or have limitations or exclusions-make sure you compare travel insurance offerings!

Travel insurance can be purchased in many ways:

Package Policies

These policies are designed for travellers going on a single trip. They can be used for Cruises, Tours, Flights, House Rentals and a world of other trip types.

All offer Trip Cancellation, Baggage, Medical, Dental, Emergency Evacuation, 24 Hour Traveler Assistance, Baggage Delay, Travel Delay, and Accidental Death Coverages.

Some policies also have options for Collision/Damage coverage for rented cars, Flight Insurance (a form of Accidental Death Coverage while flying only) and added Emergency Evacuation insurance.

Some package policies include coverage for children at no extra charge.

Travel Medical Policies

These policies provide Medical Coverage for travel up to a year in length. Some include coverage for evacuation, flight, accidental death, limited baggage, and trip interruption.

These policies may supplement what your province or state, or existing insurance may pay. Premiums are based upon age, time away and medical condition of the applicant.

Multi-Trip Annual Medical Policies

Policies provide medical cover for multiple trips throughout the year.

Medical Evacuation Policies

These policies provide protection for travellers in the event they become seriously ill or injured while traveling.

They provide emergency medical evacuation to the nearest appropriate care facility or on some plans, to your hospital of choice.

They also include coverage for repatriation of mortal remains.

Flight Accident Policies

Policies are designed for travellers going on a single trip and provide coverage for Flight Accident Insurance (a form of Accidental Death Coverage while flying).

Questions to ask about Travel Insurance

Here are just a few of the most common questions I am asked:

- Am I covered by my travel insurance if my mother, who is not traveling with me, develops pneumonia and I have to cancel my trip to care for her?
- If I have a heart condition and am on continuing medication, will I be covered under the pre-existing conditions exclusion if my heart condition flares up during my vacation?
- Is there anything else I should know about the pre-existing conditions exclusion?
- If my pet becomes ill and I have to cancel my trip as a result, will I be covered by travel insurance policy?

View more travel insurance frequently asked questions!

A personal story...

This one is very close to home. It was my father!

My father chose a winter vacation in California. He purchased what he thought was adequete travel insurance for himself and my mother.

I went down to join them for a short period and three days after my arrival my father had a heart attack!

Suffice it to say we did get him to the hospital in time and he did recover after two weeks in intensive care! The cost was $100,000.

The surprise-the insurance policy he chose had restrictions on pre-existing conditions and he was not covered!

Please take time to compare...

Please take time to understand travel insurance and NEVER leave home without it!

Many cruise lines offer their own travel insurance policies with various forms of coverage.

One way to compare travel insurance offerings on-line is at:

InsureMyTrip.com. This company highly integrated with its partner insurance companies and has over sixty travel insurance plans to choose from..

As soon as your order is placed on their site, the insurance company databases are updated in real time and a confirming email is sent to you outlining your coverages.

The policy is mailed to you directly from the insurance company immediately thereafter.

According to the company this level of integration is not available anywhere else and has been paramount to their success.

For your protection, please make sure you are properly insured. Cancellation/Trip Interruption insurance must be purchased at time you deposit your vacation but Medical insurance usually can be purchased up until time of departure.

Don't take a risk!

Demonstration Project for Enhancing Security and Safety by Using an Independent Law Enforcement Officer or "Sea Marshal" on Each Ship.

Project Safe Cruise

A Demonstration project for enhancing security and safety procedures on cruise ships by having an independent law enforcement officer or "Sea Marshal" on each ship.

Project Safe Cruise will consist of a one week cruise in 2007 that will showcase the type of security and safety system that could be put in place by cruise lines to insure the safety and security of all passengers and crew and at the same time preserve evidence of crimes and/or accidents. On this Maiden Voyage we hope to offer short entertaining seminars and panel discussions by various security experts and perhaps some celebrities, on missing persons, identity theft, date rape, kidnapping, teenage and young child safety, security, and general health and safety issues while in other countries or at sea, investment scams for seniors, how to buy art work, etc.

We must focus on a positive solution that will benefit all parties involved and prevent future tragedies. With the help of former FBI and NYPD law enforcement experts, we ask you to do this as a way of honouring the 61 people who have gone overboard since the year 2000. Most of whom are presumed dead.

We believe the cruise lines are able and are in fact obligated to invest in a project that could pay dividends to all parties involved in the cruise ship safety and security issue, including the cruise companies and related business, victims and their families, government oversight agencies, and all future passengers and crew. We acknowledge that cruise lines have security and safety procedures in place. However, we also believe that there is an urgent need to enhance them for the following reasons: *The number of people overboard has increased significantly in the last five years. *A common element of each disappearance and in other cruise ship crimes and safety incidents is the lack of good evidence and/or successful investigations. Few if any crimes have been solved or prosecuted. *Overnight cruise ship operators carry close to ten million passengers on over 90 ships that call on hundreds of ports worldwide. *While on a cruise vacation, American citizens have a right to have an independent law enforcement person on each ship so they can immediately report crimes and safety problems.

It has been suggested that cruise ships are small floating cities. They should have law enforcement available just the same as small cities, and not just private security personnel. The need for enhanced security will grow over time as the number of "first time" passengers increase. They, especially, should be informed about the rights and protections available to them while on the open sea or on foreign soil. There currently are no reliable, in-depth, or up-to-date statistics on the number, type and frequency of cruise crimes and accidents among passengers or crew. These must be developed so that trends can be monitored and improvements put in place to reduce the number of tragedies.

There are growing threats in the form of bird flu and other illnesses, terrorist threats, natural disasters, fires, equipment failures, and pirate attacks that underscore the need to have personnel in place that have the expertise, the experience, the training and authority to act, communicate, investigate, and monitor incidents quickly and effectively. Mysteries may be good for books, movies, and television but they offer no comfort or closure to those who have lost loved ones. We offer our support and prayers to all families as they struggle to resolve the tragedies of the past. Although we will never be able to prevent all deaths, crimes, or safety incidents that occur on cruise vacations, we can make a difference. And we can at least put a system in place that offers assurance to each and every American passenger that everything humanly possible was done to protect and/or save his or her friends and family while on a cruise vacation. A brief description of Project Safe Cruise is provided below.

Project Safe Cruise could be a win-win solution for parties on both sides of the cruise ship safety issue. Look at it as an outside the box idea, a partnership if you will. Even though focusing on prevention is less glamorous and less dramatic than solving crimes, it is more efficient and gratifying, especially if we save one life or reduce the suffering for one family. We must work with the cruise lines to set up a cadre of independent contract employees that would act as security ombudsman (Sea Marshals) who would–

- Know how to deal with a terrorist situations, fires, natural disasters, equipment failures, crime, and health and safety emergencies. Develop and maintain a security and safety action plan for each ship, which would provide guidelines and procedures to follow for different threats and incidents. Have the authority to lock down crime scenes and secure the chain of evidence.
- Help passengers file complaints with local authorities when there is theft or assaults, missing persons. Act as liaison and reviewer of safe and unsafe shore excursion operators, establishments and neighbourhoods at all ports of call.
- Maintain and inspect fire/safety/security equipment on ships and on piers.

- Develop working relationships with global and local authorities for recent threats, incidents, watch lists, etc. Be Prepared.
- Act as a trusted, friendly cop on the beat providing peace of mind for customers and crewmembers. Participate in regular training sessions and network with other Sea Marshals at quarterly conferences, sharing incidents, problems, solutions, and heads up items on a searchable intranet web-based database.
- Monitor passengers, crewmembers and related contractors for suspicious activity.
- Learn how to handle various medical emergencies including contagious disease and contamination.
- File immediate and periodic reports with the FBI, Coast Guard and other appropriate authorities as outlined in the CLASS Act.
- Develop and keep up to date various training methodologies for passengers and crewmembers.

Have available a means of communicating directly with law enforcement and emergency service personnel, such a satellite phones.

Maintain crime scene equipment including DNA kits, a safe for effective storage of evidence, temporary holding rooms, cameras, etc.

Have centralized independent legal staff available for consultation 24 hours a day when local authorities are not available.

Monitor whether was crewmembers are applying responsible service of alcohol, appropriate security measures, and not serving minors.

Deal immediately and directly with the FBI, Coast Guard, and other agencies and organizations in all ports of call.

Provide ongoing security and safety training to crew and travel agents and organizers including shore excursion groups.

Present short entertaining presentations outlining safety and security guidelines and resources for all passengers, including the design of a laminated security overview card for each cabin, similar to the ones used by air lines before each flight. Monitor training given to crew members and contract employees to verify they are fully trained and competent before reporting for duty, including surprise inspections and safety/security drills.

Record and maintain a database of safety/security/crime incidents and health inspections and develop some analysis of current procedures and conduct of parties involved and recommend improvements for all.

Based on information from Internationalcruisevictims.org, we could develop agreements with the countries where the ships are flagged and major ports of call to temporarily deputize the onboard Sea Marshals, so they can consult, cooperate with, and act on behalf of the relevant authorities until their permanent staff could arrive and take over. They could stabilize the situation and begin investigations immediately. They would preserve evidence, maintain crime scenes, and interview crew members and passengers in a timely manner. The cadre could be staffed by ex-military and ex-law-enforcement personnel who would get salary and/or some free cruise or vacation packages for their work, kind of like how many campgrounds have frequent campers act as hosts. What works good on land should work good on the water including a Neighbourhood Watch or Guardian Angel type system that could include frequent customers, crew members, contract employees, travel agents, etc. Design an Amber Alert system for use on ships, when passengers go missing. The use of this and an independent cadre perhaps named Sea Marshals would provide the most important factor or element for preventing and solving crimes: TIME. We sent information for a cruise ship security plan to the Subcommittee on National Security, Emerging Threats, and International Relations. Congressman Shays' subcommittee has introduced the "Shays/Maloney Cruise Line Accurate Safety Statistics Act"

Cruise Ships Embark on Integration

Cruise ships offer all the above, along with food, lodging and transportation, for the ultimate travel experience.

Cruise ships promise the allure of exotic locales, which are reached by crossing isolated bodies of water. Any mishap, be it weather, equipment failure or pirates, would doom the ship. As tourism booms, cruise ships launch cutting-edge solutions to preserve the lives of passengers and crewmembers.

These floating cities carry anywhere from 3,000 to 10,000 passengers, making security an utmost concern. Cruise ship operators need to keep passengers happy and out of harm's way. Both are tall orders, complicated by the ship's numerous decks and the large onboard population.

IP has boarded the newest cruise ships, such as the world's biggest Oasis of the Seas. While greenfield applications have built-in network infrastructure, retrofitting older ships with cramped quarters is trickier.

Convergence has clear benefits, but issues lurk beneath the surface. If a ship-wide network crashes, it could bring down all integrated systems, leaving the ship with no communications, security or life safety. The risk of "overintegration" requires subsystems to have stand-alone functionality, should the network go down.

As tourism booms, cruise ships launch cutting-edge solutions to preserve the lives of passengers and crewmembers. Cruise ships allow passengers to get away from the stresses of everyday life. The idyllic seafaring environment provides relaxation, but also presents unique security concerns.

A cruise ship covers all traveler needs, so guests can kick back and relax. An estimated 16.8 million people took cruises in 2008, a significant increase from about 900,000 passengers in 1983.

To meet demand, cruise lines are ramping up production. At least 15 vessels will launch in 2010, each worth about US$250 million to $1 billion, making cruise tourism a multibillion-dollar industry.

The boom in cruises is accompanied by increased regulation. Mandates, such as the Safety of Life at Sea Convention for ships, ports and government agencies, require them to "detect security threats and take preventative measures against security incidents affecting ships or port facilities used in international trade."

With a cruise ship resembling a small city at sea, security is a top priority. "The largest cruise companies generally have well-developed internal organizations, where safety and security sit as a key corporate department that is focused on continually assessing and addressing these risks," said David Fairnie, Global Ports Solutions Director for G4S. "The solutions they apply to prevent risk from materializing take the form of combining people, process and technology to mitigate the risk — never one in isolation."

Cruise ship security has parallels to the hospitality market, but the ever-changing marine environment introduces unseen issues. "There's an entire set of circumstances that are different when sailing on water in the middle of the ocean, compared to being on land with access," said Bob Swoope, VP of Sales for Alarm Lock Systems.

Terrorism

Deliberate organized attacks against densely packed cruise ships loom as an all-too-real threat. While there have been few attacks on passenger ships, cruise ships are mindful of potential hazards, such as hijacking or dirty bombs.

Cruise ships sail with detection equipment onboard expressly for screening. "Explosives, weapons and general contraband are three threats," said Brook Miller, VP of Smiths Detection. "Things like explosive devices, explosive components, guns or knives are not useful other than being dangerous things. Those threats are addressed in the screening process."

Different ports of call present varying risk levels. "Cruise ships might dock at uncontrolled or unfriendly environments, and may become easy targets for terrorist attacks," said Cosimo Malesci, cofounder and VP of Marketing for Fluidmesh. "If you take a cruise ship to the Miami harbour and blow it up, it can cause big damage."

Pirates

Pirates are among the most colourful characters at sea, embodied by Captain Hook and Jack Sparrow. Their real-life counterparts have evolved as more menacing predators, wielding automatic weapons and speedy getaway boats.

Pirates pose the biggest risk to shipping vessels or tankers, but affect passenger vessels. "Piratical attacks off the Horn of Africa constitute a threat to the lives and welfare of the citizens and seafarers of many nations," wrote the National Security Council in a 2008 report. "Nearly 12 percent of the world's petroleum passes through the Gulf of Aden, which is one of the world's most important waterways."

The attacks come at a steep price. "Shipping interests typically pay ransoms in cash ranging from $500,000 to $2 million, with the overall income from piracy ransoms estimated to exceed $30 million in 2008," read the National Security Council report.

This increased risk has resulted in more security. "In the last 18 months, there has been a rise in investment in training people and equipping vessels to better handle the risks posed by piracy," Fairnie said.

Anti-piracy procedures include crisis response to secure the vessel from piratical boarding attempts; evacuation or passenger management; and use of nonlethal sonic weapons to disrupt or deter attempts to board. "However, many companies have decided not to sail through high-risk areas in order to mitigate the risks of piracy attacks," Fairnie said.

In the April 2009 hijacking of the Maersk Alabama, the crew used its anti-piracy training to stop the Somali pirates from controlling the freighter. "A video surveillance camera system alone is not going to prevent pirates from boarding a ship, but an integrated security system incorporating radar, thermal imaging, communications and defensive countermeasures just might give the ship's crew a fighting chance," said James Ionson, CEO of Oncam Global.

For the Maersk Alabama, the pirates kidnapped the captain but did not secure a ransom. "The US Navy used cameras onboard its P3 Orion and Scan Eagle unmanned aerial vehicle to identify how many pirates were onboard and what armament they had, and to monitor the ship's progress toward port," Ionson said. "Snipers took out the three hijackers, leading to the dramatic rescue of Captain Phelps."

Smuggling

Passengers are welcome onboard ships, but stowaways or controlled substances are not. Maintaining this delicate balance requires careful oversight and technology. "At many different ports of call, the cruise ships need to make sure they are not transporting contraband, weapons or threat items from one place to the next," Miller said. "It's like a movable airport, with a high concentration of people and equipment in one very small area."

The process of embarking and disembarking passengers presents the highest risk of stowaways or illegal substances being transported. "This risk is usually addressed by a clear management process which incorporates controlled passenger routing, documentation checks and profiling — combined with security technology such as screening of baggage, stores and passengers — using walk-through and handheld metal detectors and X-ray systems, operated by trained security personnel," Fairnie said. The boarding phase is coordinated by ship personnel, along with outsourced security professionals landside, he said.

Environmental Hazards

The high seas present challenges to seafarers. Unexpected natural disasters, such as ice and flash flooding, must be accounted for, said Johannes Rietschel, founder and CEO of Barix.

The ship it self is unlike a land-based installation. "Some unique concerns cruise ships have are highly integrated systems, severe environmental factors, challenging cable runs and cramped equipment room spaces," said Rodger von Kries, VP of Technomad.

Limited ship space results in a tight fit for passengers and equipment. "Closed compartments and watertight doors pose challenges for installation," Malesci said.

Cruise ships must communicate clear messages to passengers throughout the vessel. "In our view, zoning is especially important on a vessel where you have many decks, corridors and limited visibility," Rietschel said. "If something rocks the boat, there needs to be a way to get precise instructions out quickly, and these instructions might differ for various parts of the vessel."

In the event of system malfunction, a cruise ship features multiple backup systems. "It's similar to how hospitals run power," Malesci said. "Usually on a ship, everything is doubled for power, water or electrical systems."

Close Quarters

Passenger ships cater to a distinct demographic. "Cruise ships tend to attract an older clientele, many of whom have existing health problems, and so crew must be able to respond rapidly to heart attacks, strokes or even ordinary seasickness with appropriate medical treatments," Ionson said.

A real threat for an onboard population would be epidemics. "Some threats would be to health environments," Swoope said. "It could be illnesses."

Some threats are universal, whether on land or at sea. "Ship operators are concerned with occurrences of such shipboard incidents as missing persons, accidents involving passengers and crew (legitimate and fraudulent), and theft," Ionson said.

Cruise ship passengers could become targets in foreign countries. "In the Bahamas, the harbour had problems of thieves attacking tourists leaving cruise ships," Malesci said. "With cameras, the police could keep tabs on the thieves."

Accidents aboard cruise ships could result in lawsuits. "You have liability for cruise ship operators," said Erik Laurence, VP of North Asia, Verint Systems. "If someone claims they fell and hurt themselves and blames the operator, reviewing the video can help in many cases."

Cruise ships watch out for passenger safety. "For areas around decking or the pool, you can detect if there's a hazard somewhere on the ship and protect the owners from being sued," said Brian Lane, Senior Product Manager for Spectra, Pelco (a Schneider Electric company).

A cruise ship faces hazards common to hospitality and airports, along with unique environmental challenges. While being at sea presents danger, cruise ships take safety and security seriously by deploying holistic solutions.

Cruise Ship Passengers

Cruising is statistically the safest form of transport in the travel industry and takes place under strict security conditions. But there are a few things you should be aware of. This advice has been prepared in association with the Passenger Shipping

Association (PSA). As with all information provided by third parties, the FCO is not responsible for its accuracy.

How do cruise ships plan for emergencies?

- safety drills are mandatory and details about these are found on daily newsheets or TV programmes provided by the crew
- it's common practice for safety drills to take place either immediately before departure or shortly afterwards.
- familiarise yourself with the safety drill information as soon as you reach your cabin.
- during your cruise you might see and hear a number of drills for the crew where they respond to a variety of 'emergency' situations.

What kind of security can I expect on board cruise ships and the ports they visit?

- most cruise ships have CCTV, which can be used to assist with onboard investigations
- security checks have been enhanced and are a similar standard to those of airports
- when the ship is in dock you'll be security checked whenever you disembark and return to the ship.
- passengers are normally given ID cards on check-in, which you must show to enter and exit the ship.

Cruise lines, ships and the ports they visit are required by law to have appropriate security plans.

The deployment of security staff is usually discreet so you may never notice them but there should be trained security personnel on board at all times.

What happens if I am the victim of a crime?

You should report it to a crew or staff member. Security personnel should then inform the appropriate authorities. Who these are will depend on where the ship is, when the crime is committed and the nationality of those involved.

Significant crimes against British nationals on any ship can in certain circumstances be reported to UK police and may be investigated even though they occurred outside the territory of the UK.

A crime may also be reported to the authorities in the port/ country in which the ship was docked (or was headed) when the crime was committed with the result that local law enforcement agencies may also be involved in the investigation.

Lost or Stolen Passports

If your passport is lost or stolen during the course of a cruise you may need to obtain a replacement passport from the nearest British Embassy, High Commission or Consulate before you can continue your cruise. You may not be able to continue your cruise if a replacement passport is not available before the ship sails. Clarify this situation with your cruise company before travelling, especially about the arrangements they have in place for you to re-join the ship at a later stage.

How can relatives/friends contact me in an emergency?

- At the time of booking you should be given printed information on how to contact the ship by telephone, fax or email -leave this with friends or relatives
- leave a copy of your travel itinerary with family or friends before you travel
- almost all ships have satellite telephones in cabins, and many also have internet cafes
- ships often have wi-fi technology so that you can use your own laptops in your cabin or around the ship
- some ships enable you to use your mobile while the ship is out at sea as well as in port.

Visiting Multiple Countries-Visas

Cruise lines may be able to organise group visas for those going on organised tours in certain countries. Sometimes, the ship can acquire visas for you as it proceeds towards the port, and this cost is typically added to your shipboard account. For other destinations, you might have to obtain your own individual visas even if you don't intend going ashore- check this at the time of booking with either the travel agent or the cruise line.

Make sure you have the necessary documentation in place if the country where you start the cruise requires you to have a visa – there's more information on entry requirements in our travel advice by country.

If you're refused entry to the country and miss your cruise the cruise line will not ordinarily be required to pay you compensation.

Health Issues

Medical care is not free on cruise ships, even on British flag ships. It's essential that you take out adequate insurance cover, including cover for the cost of medical treatment or medical evacuation – in an emergency you may have to be airlifted to port if the ship is at sea.

If you're cruising within Europe ensure you have an EHIC card – but you'll still need comprehensive travel insurance as well! Your EHIC will only cover necessary treatment required during a visit, for example to see relatives, to an EEA member state where the obtaining of treatment was not the reason for your being in that country.

All cruise ships should have a qualified doctor and medical facilities on board. Although the range of facilities varies from ship to ship and line to line, the minimum standard is designed to provide emergency medical care and the capability to stabilise patients. Discuss any concerns you may have about your medical needs and the level of care available on board with your cruise company. If you require emergency medical treatment that means that you need to be treated on land you should be aware that it may not be possible for you to return to the cruise later. In these circumstances the cruise ship (or their representative in-country) should:

- ensure that you are placed in the most suitable medical facility, including liasing with your insurance company;
- arrange for you to re-join the ship at a later stage, or be returned to the UK when you are well enough to travel;
- arrange accommodation for your travelling partner/Next of Kin while you are in hospital; and
- advise you on how to transmit appropriate funds.

Ensure that you have all your documents, including your passport and insurance policy, with you and a contact list for unforeseen events. You should also remember to take with you, if possible, any items that you or your companion may need when ashore, especially if you are not able to rejoin the cruise at a later date. In an emergency you can contact the nearest British Consul.

Prescription Medicines

If you are on prescription medicine bring enough to last the duration of the cruise, as it may be difficult as well as costly to purchase these on board.

Gastrointestinal Illnesses

Given your close proximity to other passengers, and the fact that you are in a confined area, outbreaks of gastrointestinal illness do occur from time to time although incidents on cruise ships are less common than on land. Many ships are likely to ask if you have recently experienced any typical symptoms before embarkation and you may be denied entry if there are genuine concerns.

If you are diagnosed with a gastrointestinal illness on board, you may be asked to remain in your cabin until cleared by the medical staff, so as to help reduce the potential that the infection will spread further amongst passengers and crew.

Cruise ships undergo regular, unannounced inspections by UK port health authorities to help ensure they meet the statutory food safety and hygiene requirements, and if they operate in the US they also have to meet the CDC Vessel Sanitation Program standards.

Crew members should be trained to follow routine hygiene and sanitation procedures to help prevent outbreaks and deal with any that do occur.

Cruise Ship Safety and Security For Families

While there have been incidents of crime at sea, most families feel that cruising is a safe and secure experience for their kids. Many parents actually give their kids a bit more freedom to explore on their own while the ship is at sea, especially older grade schoolers and teens; after all, the cruise line controls access to the ship, and uniformed staff are almost always right around the corner.

Nonetheless, parents certainly should discuss safety issues with their child before giving them too much freedom. Obviously, they should know basics like not climbing on railings and how to respond if the ship's emergency signal sounds — all children younger than teenagers have to wear an identification bracelet during the cruise that lists their muster station, so youth counsellors can quickly identify where kids need to go to meet their parents. In any event, kids should also be ready to approach the ship's staff if they need help, recognizing them by their uniforms and name tags. Also, children should know that they can use the house phones where they can call the front desk or your cabin in the event of a personal emergency. Regardless of your child's age and gender, there are a few other issues that should be discussed as well as a some groundrules that should be observed.

Most children probably already understand this, but it bears repeating: They should never go into anyone's cabin without parental approval, and on the elevators, which are kid magnets, they should have a strategy if someone makes them feel uncomfortable. As a family, you all should walk the route from the cabin to the youth room and other key locations together so that it's all familiar and nobody gets lost. The lower decks where the crew resides are off limits to all guests, and children should know this.

The youth programs require that children are signed in and out of their youth programs, and while the age varies from line to line, children under 9 or 8 years must be picked up and signed out by a parent, often with a photo ID. Older than that to early teens, the parental sign-out policy is customized at the start of the cruise. Communication is key. Everybody should be clear about what time they will meet up again, and where. Parents should encourage children to leave notes in the cabin describing their plans; walkie-talkie radios can be quite useful with teens who are allowed to move about as they wish.

While you may allow children some free rein on the ship, they should not stray from parental supervision during port visits. The only exception would be a shore excursion for kids, supervised by ship staff, with the children understanding that they cannot separate from the group. Adults, who should also know how to respond, can be targets anywhere tourists flock; let's not tempt the would-be criminals with our children. Instead, use the port visits to explore and to experience a new place ports together, as a family.

Disney Safety and Security

Disney Safety and Security is a The Walt Disney Company division responsible for workplace and guest safety, security and loss prevention at Disney resorts, theme parks, studios, corporate offices, Disney Cruise Line. This division also defines policies and programs for Disney's online presence and consumer products.

Disney company policy does not allow employees to carry weapons, including security personnel. Disney is exempted from Florida law which allows employees to carry weapons in their cars if they have a concealed weapons permit.

Resorts and Cruise Line

Disney Safety and Security oversees the hotel and cruise line security, lifeguard hiring and training, and food safety program.

Disney trains and employs 1,200 lifeguards to monitor pools located at its resorts, on its ships, and at the beaches of Castaway Cay. Lifeguards are trained in water rescue, CPR, basic first aid, oxygen administration and the use of Automated external defibrillators. Disney lifeguards have contributed to lifeguard training manuals and programs used nationwide.

Disney's food safety program provides laboratory and on-site testing of food products throughout the delivery, preparation and service process.

The Walt Disney World Safety and Health Department was recognized in 2001 for its food safety programs in place at resorts, theme parks and the cruise line by the International Association for Food Protection receiving the Black Pearl award. The director there received the 2007 Lifetime Achievement Award for Leadership in Food Safety from NSF International.

Theme Parks

Disney's security personnel are generally dressed in typical security guard uniforms though some of the personnel are dressed as tourists in Plain clothes. Since September 11, 2001, uniformed security has been stationed outside each Disney park in Florida to search guests bags as they enter the parks. Ticket Sellers search bags at the California Parks. This also now occurs at Tokyo Disney Resort, Disneyland Resort Paris, and Hong Kong Disneyland Resort, outside the main entrance of all the parks.

Walt Disney World Resort

The land where Walt Disney World resides is part of the Reedy Creek Improvement District (RCID), a governing jurisdiction created in 1967 by the State of Florida at the request of Disney. RCID provides 911 services, fire, environmental protection, building code enforcement, utilities and road maintenance but does not provide law enforcement services. The approximately 800 security staff are instead considered employees of the Walt Disney Company. Arrests and citations are issued by the Florida Highway Patrol along with the Orange County and Osceola County sheriffs deputies who patrol the roads. Disney security does maintain a fleet of security vans equipped flares, traffic cones, and chalk commonly used by police officers. These security personnel are charged with traffic control by the RCID and may only issue personnel violation notices to Disney and RCID employees, not the general public.

Despite the appearance of the uniformed security personnel, they are not considered a legal law enforcement agency. Disney and The Reedy Creek Improvement District were sued for access to Disney Security records by Bob and Kathy Sipkema following the death of their son at the resort in 1994. The court characterized Disney security as a "night watchman" service not a law enforcement agency and was not subject to Florida's open records laws. An appeals court later upheld the lower courts ruling.

Disney Security personnel in Florida have investigated traffic accidents and issued accident reports. The forms used by Disney Security may be confused with official, government forms by some.

The Orange County Sheriff maintains an office on Disney property, but this is primarily to process guests accused of shoplifting by Disney security personnel.

A Qualitative Study of Victimization and Legal Issues Relevant to Cruise Ships

Cruise Line History

Whether it was the disaster in 1912, or the 1997 block-bluster hit, the Titanic is perhaps the most documented and well-known cruise of all time. The cruise industry today however, is not associated with icebergs and disasters but with large ships full of gaming activities on the way to beautiful destinations. "Cruise lines, diverging from the industry's earlier affluence and glamour, began to market more affordable vacation cruises with a 'fun ship' image". This new image has proven to be very prosperous for the industry.

Cruise line companies have enjoyed the monetary benefits of their success, and continually monitor the growth of the overall industry. Over 12.5 million people took a cruise in 2007, as reported by the Cruise Line International Association (CLIA). The total number of passengers recorded in 1990 was 3.7 million. Additionally, the cruise industry has witnessed an average growth rate of 7.4% annually. In 2004 U.S. citizens accounted for 76 percent of the global market share, and has gradually declined since, reporting 73 percent in 2006. The CLIA remains optimistic, reporting that 34 new ships are contracted or planned to be added to the North American fleet from 2008 through the end of 2012. This will accompany the 185 cruise ships already in existence. The International Cruise Line Association is composed of 23 of the major cruise lines, representing 97 percent of the cruise capacity in North America. CLIA was formed in 1975 and merged with the International Council of Cruise Lines in 2006.

Given the current state of U.S. economic affairs it is impressive that cruise lines are experiencing such growth in the leisure market. Even with the decline in U.S. passengers the industry continues to grow, counteracting the sinking economy by focusing on foreign markets. Aside from the revenue generated by U.S. passengers, cruise lines are independent of the U.S. economy. Even though nearly 75% of passengers are U.S. citizens "cruise line corporations and their ships are not traditionally American-owned or registered..." Thus, cruise ships are not affected by changing laws inflicted upon U.S. corporations. In reality, cruise lines are still flourishing while the U.S. economy suffers.

Cruise line companies are not concerned about increasing minimum wage, rising insurance premiums, or higher corporate taxes. Cruise lines escape federal taxes and labour laws by registering their corporations and vessels in foreign countries, i.e. Panama, Liberia, and the Bahamas. In fact, employees of cruise lines are often mistreated due to lackadaisical labour laws. Worst of all, employees will find little to no recourse pursuing litigation. Likewise, a U.S. citizen passenger faces the same predicament.

For example:

"A person who goes out to his local neighbourhood bar and becomes so intoxicated that he is unaware of his surroundings (and) subjects himself to any number of situations-including criminal acts by other persons. While that person may be able to sue the bar civilly if his drunkenness results in his own injury, if the same events happen on a cruise ship, he may have no recourse".

Employees and passengers may have difficulty litigating claims due to the fact that "international maritime law applies to the cruise industry because cruise ships sail in the waters of many countries, as well as in the waters of no country, otherwise known as the high seas".

Before immersing into the ramifications of federal maritime law, it is important to consider the context and history in which this law was developed. "...Consumers should be aware that the cruise ship's duties and liabilities are governed not by modern, consumer-oriented common and statutory law, but by nineteenth-century legal principles, the purpose being to insulate cruise lines from the legitimate claims of passengers".

Courts today base their decisions on federal statutes enacted almost a century ago. Essentially, these statutes are the reason cruise lines have been able to dismiss claims brought against them from both passengers and employees. "The antiquity of maritime law was made manifest in Barbetta v. S.S. Bermuda Star [italics added], a case approving the rule that 'shipowners are ... not vicariously liable for the negligence of ship's doctors in treating passengers".

However, in August of 2003, the Florida Third Circuit Court of Appeal rejected Barbetta and held that a ship's doctor's negligence should be imputed to a cruise line. In February of 2007, the Florida Supreme Court quashed the decision of the district court and held that the ship owner is not vicariously liable under the theory of respondeat superior for the medical negligence of the shipboard physician (Carnival Corporation v. Carlisle, 2007). In most cases, however, the Shipping Act of 1984 (46 USCS § 30511) has been applied, relieving cruise lines of liability for the negligent acts of its cruise ship staff (e.g., a shipboard doctor).

"Most importantly, the aggrieved passenger and his or her attorney should be aware that the passenger's rights and remedies are governed by maritime law which in many important respects is very different from common law". Unlike common law, maritime law enables cruise lines to significantly limit their personal liability, as well as a passenger's rights and remedies. Worst of all, the ticket-passenger contracts and the limiting-liability clauses within are enforced in most U.S. courts.

Flags of Convenience

Cruise lines have been circumventing U.S. statutes and regulations since as early as the 1920's. "A number of U.S. vessels, namely the cruise liners the M/V RELIANCE and the M/V RESOLUTE, were 're-flagged' in Panama in order to circumvent the U.S. law banning the sale of alcohol on U.S. registered ships (during prohibition)". The International Maritime Organization (IMO), to which the United States is a member, "requires all ships engaged in international trade to have a country of registry in order to sail in international waters. A ship is considered the territory of the country in which it is registered..." The verbiage of this clause is particularly important as it states "a country" and could pertain to any country, developed or undeveloped. Even more importantly Tomlinson states that the ship is considered the territory of its country, thus making them subject to that country's laws and regulations.

"By opting to re-flag in a new nation, a vessel owner becomes subject to the safety, labour, and environmental codes of that nation. Thus, those nations whose open registries have become the most popular also tend to be those who possess the most lax labour, safety, and environmental codes".

A vessel's country of registration is commonly referred to as the "flag of convenience (FOC)." Flagging a ship under a foreign flag for the convenience of the cruise line is nothing new, nor is it rare. The majority of cruise ships today are registered to Panama, Liberia, or the Bahamas. It is important to pay close attention as many vessels within the same fleet are often registered to different countries. Carnival Corporation, for example, has flagged their cruise vessel Celebration under Panama and Destiny under the Bahamas. Cruise lines often avoid drawing attention to the FOC of by using the term "headquartered in Miami, Florida." It is important to understand that while the majority of these cruise lines have their headquarters in Miami, they are not registered in the U.S. Thus, U.S. laws do not apply and passengers are at the mercy of maritime law.

Consequences of the Flag of Convenience

The legal rights and remedies of U.S. passengers are greatly inhibited due to flags of convenience, partly because ticket-passenger contracts are written under the registered country's law and can include limited-liability clauses. Under maritime law, such clauses are acceptable and enforced in court. Regardless of litigation locale, the majority of courts apply federal maritime law because crime while at sea falls under the admiralty jurisdiction.

"Most passengers do not realize that a cruise ticket forms a binding contract between the passenger and the cruise line. These passenger-ticket contracts have been heavily litigated because they contain limited-liability clauses such as forum-selection clauses, choice-of-law provisions, and notice-requirement clauses that limit the rights of passengers".

"Forum-selection clauses, provide that any controversy arising out of the cruise contract are to be litigated, if at all, in a certain jurisdiction to the exclusion of all others". The most important case to date in regards to forum-selection clauses is Carnival

Cruise Lines v. Shute. In 1991, the "United State Supreme Court made several rulings that remain the law today" (Porter, 2006, p. 601). The forum-selection clause, as printed in the ticket-passenger contract, mandated that Shute file claims and bring lawsuit against Carnival Cruise Lines only in the state of Florida. The Shutes argued that the forum-selection clause "was not the product of negotiation, and enforcement effectively would deprive [them] of their day in court.' If enforced, the clause violated the Limitation of Vessel Owner's Liability Act, codified in 46 U.S.C. § app. 183c". The court dismissed the case based on precedent set in Bremen v. Zapata Off-Shore Company (1972).

In doing so, the court solidified the enforcement of forum-selection clauses on three accounts:

"First, the court observed that because a cruise line carries passengers from many locales, it could be subject to litigation in several different forums, and would thus have a 'special interest in limiting the for[ums] in which it potentially could be subject to suit.' n59 Second, the court noted that such clauses have the practical effect of dispelling any confusion about where suits must be brought and defended, as a result, spare 'the time and expense of pretrial motions to determine the correct forum and [conserve] judicial resources that otherwise would be devoted to deciding those motions.' Third, the Court opined that 'passengers who purchase tickets containing a forum clause... benefit in the form of reduced fares reflecting the savings that the cruise line enjoys by limiting the for[ums] in which it may be sued".

The federal courts have adapted a test for measuring the enforceability of forum-selection clauses, known as the two-prong test. In order for forum-selection clauses to be enforceable, the terms of the contract must be reasonably communicated to the passenger and the clause must be fundamentally fair. There are some exceptions in which forum-selection clauses are not enforced. However, for the most part, the rule solidified by Shute sustains, and passengers must litigate at the location stated in the ticket contract.

Shute additionally brought claim that forum-selection clauses were unenforceable on the basis of bad faith (the cruise line was trying to dissuade passengers from pursuing legitimate claims), but the Court ruled that no evidence existed. Mostly in part because Carnival Cruise Line is headquartered in Miami and the majority of their vessels deport from Florida, thus giving them motive to litigate in Florida.

Arguing Forum-Selection Clauses

"Plaintiffs bear a heavy burden of proof in order to invalidate a forum-selection clause on these grounds, and concerns about cost, distance, disability, or age are not considered tantamount to depriving litigants of their day in court". Although most courts uphold forum-selection clauses, there are a few exceptional circumstances in which they rule against them.

"According to the 'two-prong test' the passenger's inability to reject the ticket-passenger contract without sacrificing part or all of their monetary investment is unfair and thus, unenforceable. U.S. courts validate contracts if: (1) they are reasonable; (2) the contract is not fraudulently induced; and (3) the passenger is given the opportunity to reject that contract". If the cruise line "sent (a cruise ticket) at a time when (the passenger) could not conceivably have cancelled without avoiding a penalty", the court may rule the contract is invalid.

In order for a contract to be considered fair, a passenger must be able to negotiate or disagree to the terms without consequence. As litigated in Schaff v. Sun Line Cruise, Inc. (1998), the plaintiff did not have the opportunity to reject the ticket contract without forfeiting the entire price of the cruise and therefore "the forum-selection clause as it applied to the plaintiff was fundamentally unfair". If the passenger receives the contract and still has time to cancel their cruise provided a full refund, then the court will hold the forum-selection clause as enforceable (Vega-Perez v. Carnival Cruise Lines, Inc., 2005). "Some courts have agreed that imposition of a cancellation penalty means that the notice was inadequate rendering the forum selection clause unenforceable while other courts have rejected this concept".

Another instance where the forum-selection clause was not enforced is in the case of Walker v. Carnival Cruise Lines (2008). Walker purchased a cruise ticket from a travel agent who had assured him (along with the cruise line) that both his room and the cruise ship would be disabled accessible. After boarding, he found that neither his room nor the ship itself was in compliance with the ADA (American Disabilities Act, 1990). The court first dismissed the claim and moved the case from California to Florida, on account of the forum-selection clause. Later, the court decided not to enforce the forum-selection clause for two reasons. "First, the plaintiffs' physical disabilities and economic constraints were so severe that, in combination, they would preclude plaintiffs from having their day in court. Second, was the fact that the plaintiffs were seeking to vindicate important civil rights".

Courts uphold forum-selection clauses so often that cruise lines "have threatened to move for sanctions against plaintiffs bringing suit in a forum other than that specified in the contract, n67 or to counterclaim against plaintiffs for breaching the contract of passage by not bringing suit in the specified forum".

Choice-of-Law Provisions

Similar to forum-selection clauses, the ticket-passenger contract dictates which applicable laws will be used if claims are brought against the cruise line.

For instance in the case of Mansoor v. M/V Zaandam (2006) the court ruled that:

"This contract and its interpretation shall... be governed by and construed in accordance with the general maritime law of the United States; to the extent such maritime law is not applicable, it shall be governed by and construed in accordance with the laws of the State of Washington (U.S.A.)".

For the most part, maritime law applies to any claims brought against a cruise line for incidents that occur while at sea.

In the absence of relevant maritime law, courts may apply state law, so long as the application of state law "does not frustrate the national interest in having uniformity in admiralty law".

An instance in which state law replaced the absence of maritime law was in the case of Jane Doe v. Celebrity Cruises, Zenith Shipping Corp., et al. (2004):

"The general rule is that the assault and rape of a plaintiff by the defendant's employee is considered to be outside the scope of employment and, therefore, insufficient to impose vicarious liability on the employer (citations omitted). However, like other states, Florida recognizes an exception for common carriers (citations omitted). For example, (citations omitted), a Florida court held that an ambulance service was vicariously liable for an ambulance attendant's sexual assault of a passenger because the assault occurred en route to the hospital. The court explained that an 'established exception to the general rule [of nonliability] is where the employer is a common carrier for hire to the public, and the tort or attack is by an employee upon a passenger while the contract for transport is being accomplished (citations omitted)."

Although state laws may favour passengers, cruise lines are aware of the laws that apply in each state and are likely to have well established defences. Furthermore, cruise lines purposefully select states in their forum-selection clauses that will rule in their benefit. Certain states such as: Idaho, Montana, North Carolina, Alabama, and Georgia do not enforce forum-selection clauses. Therefore, cruise lines will limit passengers to only bring claims in states that enforce such clauses, as well as other clauses in their favour.

Notice-Requirement Clauses

Notice-requirement clauses (also stated in the ticket-passenger contract) limit the time in which passengers can file a claim against the cruise line. "Under maritime law, plaintiffs generally are afforded a three-year statute of limitations for personal-injury claims. However, the Limitation of Vessel Owner's Liability Act (2000) allows cruise lines to establish, through passenger tickets, a one-year statute of limitations". "For nonphysical injury claims, cruise lines may impose even shorter time limitation periods such as requiring that written suit be filed within six months instead of the one year allowed for physical injury lawsuits or requiring that written claims be filed within days as opposed to months after the accident".

In the case of Johnson v. Commodore Cruise Lines, Ltd. (1995) the defendant asserted that the plaintiff failed to notify it of the claim within the constraints stated in the ticket-passenger contract. According to the contract, Johnson would have had to notify

Commodore within six months of the alleged rape and would have needed to initiate a suit within one year. Johnson brought the suit one year and seven months later, thus the court found that Johnson's claims were time-barred.

In contrast, in Deck v. American Hawaii Cruises, Inc. (1999), American Hawaii Cruises stated in their ticket contract that "other" claims must be filed within 6 months with a 15-day notice provision.

"The court concluded that such language might not reasonably alert a passenger of the time limitation applied to federal and state statutory rights, as might the use of a more inclusive term, such as 'any' instead of 'other.' n115 Thus, the court held the limitation to be unenforceable, particularly because [c]ruise passenger tickets are contracts of adhesion and as such, ambiguities in them must be construed against the carrier".

Typically, courts enforce notice-requirement clauses, unless the ticket-passenger contract is unclear in language. Or as in the case above, failed to specify which claims are time-barred. In response to these outcomes, cruise lines alter their contracts to avoid the use of ambiguous language. Thus, making it very difficult for the passenger to argue against notice-requirement clauses held within the contract.

Negligence

To prove negligence, a plaintiff must show: (1) that defendant owed plaintiff a duty; (2) that defendant breached that duty; (3) that this breach was the proximate cause of plaintiff's injury; and (4) that plaintiff suffered damages (Erin Lynn Baker v. Carnival Corporation, et al., 2006).

"Under general maritime law a shipowner owes passengers the duty of exercising reasonable care under the circumstances. Kermarec v. Compagnie Generale Transatlantique (1957). However, a shipowner is not an insurer of its passengers' safety.... There must be some failure to exercise due care before liability may be imposed. The standard of reasonable care generally requires that a carrier have actual or constructive notice of the risk-creating condition. Where it is alleged, however, that defendant created an unsafe or foreseeably hazardous condition, a plaintiff need not prove notice in order to show negligence. The question becomes whether the defendant created an unsafe or foreseeably hazardous condition".

Today, most passenger tickets include clauses that disclaim liability for problems that may occur during the cruise. In the U.S., contracts cannot dismiss claims for negligence that is wilful and wanton or gross.

However, "as with consumer contracts on dry land, instances of gross negligence and intentional misconduct may not be disclaimed by common carriers". The challenge becomes providing enough evidence to prove that carriers were negligent in their actions under maritime law.

According to the Shipping Act of 1984, a ship-owner owes to its passengers the duty of a high degree of care. The ruling of Kermarec v. Compagnie Generate Transatlantique (1957) applied the rule of reasonable care under the circumstances. The problem arises however, when laws and regulations have not been established, of which 'reasonable care' can be based upon. For example: a company in the U.S. must abide by labour laws. An employed sixteen-year-old is allowed to work for X amount of hours and requires X amount of breaks. If this employee is injured from working too many hours without a break, a breach has occurred. There was a duty of care owed; that duty was explicit, the employer ignored it, and thus is negligent.

Now, consider a sixteen-year-old employee on a cruise ship who abides by the labour laws of the flag of convenience. Without strict labour laws, the duty of care owed becomes ambiguous. Aforementioned, one must prove there was a breach of contract in order to prove negligence. An employee has a difficult time proving breach existed because the duty of care owed is not explicit. The fewer rules, regulations, and guidelines a company has, the more difficult it becomes to prove the company was negligent.

Furthermore, ticket-passenger contracts greatly limit the amount of punitive and exemplary damages one can receive. Thus, even if the passenger proves the alleged negligence, damages received are bound to the constraints of the contract and maritime law.

Medical Malpractice

Traditionally, cruise ships have not been held vicariously liable for the medical malpractice of the ship's doctor or medical staff.

In the case of Barbetta v. S.S. Bermuda Star, the Fifth Circuit stated: ..."When a carrier undertakes to employ a doctor aboard a ship for its passengers' convenience, the carrier has a duty to employ a doctor who is competent and duly qualified. If the carrier breached its duty, it is responsible for its own negligence. If the doctor is negligent in treating a passenger, however, that negligence will not be imputed to the carrier".

Fortunately, today more courts are holding cruise lines vicariously liable for the intentional torts of their employees. Nadeau v. Costley (1994) is a case on point.

As of now, there are no requirements for the training and competency of hired medical personnel onboard cruise ships. In fact, a study conducted by Feuer and Prager, which evaluated both medical facilities and medical staff onboard eleven passenger cruise liners found that "27% of doctors were not certified in advanced cardiac life support; 54% of doctors were not certified in advanced trauma life support, and 45% of doctors were not board certified in their area of practice". The extent of which cruise lines are held vicariously liable is truly determined on a case-by-case basis. As recently as 2005, the Southern District of Florida upheld Barbetta (1987) which held that cruise lines are not liable for ship doctor's malpractice".

Medical malpractice and accountability onboard cruise ships will always be a problem until something is done to change it. Laws and regulations need to be established to prevent cruise lines from hiring inadequate personnel. Regulations need to be in place to hold cruise lines accountable.

Methods

Participants

Interviews were conducted with female U.S. citizens over the age of 18 who were victims of assault while onboard a cruise ship. There were certain circumstances in which a victim could not conduct an interview. Under these circumstances, someone directly related to the victim had the opportunity to voice their story. To obtain participants, a memo was distributed, to which victims responded at will.

Materials

Interviews were conducted over the phone and participants had the option to be audio recorded. Audio recording was not necessary for the interview to take place. The interviewer conducted interviews in an isolated soundproof area with no distractions. Interviewees had the option of interviewing in any location they felt comfortable. If the interviewee decided to have the interview audio recorded, a small audio recording device was used during the conversation. The memo, consent forms, and interview questions were prepared and approved by the University of Florida Institutional Review Board prior to interviewing.

Measures

Interview questions were designed to obtain a general overview of the litigation process relevant to cruise ships. Responses were divided into two different categories: criminal and civil.

In regards to criminal, data was measured based on: attempt of conviction, the statutes/case law/regulations that affected conviction, and overall impact of the governing law.

In regards to civil cases, data was first sorted based on what type of lawsuit was filed and the governing law that applied. Cases were reviewed for allegations against the cruise line for negligence regarding a crewmembers' assault. If alleged, outcomes were reviewed to see if the cruise line was held liable for the intentional torts of their employee via respondeat superior.

For both civil and criminal cases, if admiralty was the governing law applied then cases were reviewed for the enforcement of limited-liability clauses. Cases were then reviewed for all other applicable laws specific to maritime law. All cases were reviewed for risk management procedures, focused on collection of evidence.

Procedure

After informed consent procedures, interview times and dates were arranged between interviewees and the principal investigator (Caitlin Burke). Interviews conducted lasted approximately an hour in length and followed the interview questions. Follow up interviews were not necessary for the purpose of this project. Interviews that chose to be recorded were transcribed within 9 days of the interview. Victims were at liberty to discuss what they felt comfortable following the parameters of the interview. Personal information given during interviews regarding specific details of their assault was omitted. Interview results were first compared on a case-by-case basis, and then cross-examined for similarities and differences amongst participants.

Results

Case 1: Victim and alleged assaulter were U.S. citizen passengers. The incident took place in international waters on a foreign flagged vessel. Evidence was not collected. The victim's room was sealed with security tape and guarded by cruise ship security. Victim's mother was told (by cruise line) "they are not in the business of conducting investigations." Mother brought a lawsuit against the cruise line for medical malpractice and wrongful death. Victim's mother still wants to press charges against the victim's cabin mate but lacks sufficient evidence. Maritime law governed plaintiff's claim against the defendant because the incident happened more than three nautical miles off the coast. Forum-selection, notice-requirement, and choice-of-law clauses were all enforced in court. The court ruled that the cruise line could not be held vicariously liable for the ship doctor's alleged malpractice. Under admiralty law, the Death on the High Seas Act (DOSHA) applies. Recovery under DOSHA is limited to fair compensation for the pecuniary loss sustained by the individuals for whose benefit the action is brought (46 U.S.C.S. § 30303, 1983). Non-pecuniary damages are not recoverable. As, far as outcome is concerned, the enforcement of this law severely limited the damages received. Monetary value was based on her daughter's worth prior to death (worth meaning how much she earned annually). DOSHA does not provide compensation for wrongful death or pain and suffering.

Case 2: Victim was a U.S. citizen. The alleged assaulter was a foreign crewmember employed by the cruise line at the time of the assault. The incident took place in international waters on a foreign flagged vessel. Evidence was not properly collected;

evidence personally collected by the victim was turned over to cruise ship security personnel. The victim was later told that the FBI was given all evidence. The crewmember was fired and sent to his home country immediately following the incident, thus victim has not been able to press charges against him until he is found. The victim brought a lawsuit against the cruise line for negligence in regards to several torts. The incident occurred over three nautical miles off the coast, thus, admiralty law applied. Forum-selection, notice-requirement, and choice-of-law clauses were all enforced in court. The case was eventually settled after over a year of trial without judgment. Victim stated: "I needed to redirect my focus on continuing to heal myself and getting better so that I could help people like (victim's name omitted) and other women out there. You know I just couldn't deal with (cruise line name omitted), I wasn't going to get anywhere with them! I was better to get out of that and end it… and then move on from that." Victim is not at liberty to discuss the amount she received.

Case 3: Victim and assaulter were both U.S. citizens and crewmembers. The incident took place at a U.S. port on a vessel registered to the U.S. Evidence was not collected properly and the victim was refused a rape kit and examination. Victim was "taken to a U.S. hospital were I met the U.S. police department, and they couldn't figure out who would be in jurisdiction." The victim was unable to press charges against the alleged crewmember due to lack of evidence. When questioned, the alleged assaulter claimed he never entered the victim's room. He later changed his story saying that she seduced him and that it was consensual. The FBI claimed that it could not be tried in court, because it was a he-said, she-said case due to lack of evidence. The victim later tried to file a lawsuit against the cruise line and the U.S. hospital. However, her lawyer dropped her case when she refused to sign a gag order.

Case 4: Victims were U.S. citizens. The alleged assaulter was a foreign crewmember employed by the cruise ship at the time of the assault. The incident took place in international waters on a foreign flagged vessel. Assaulter entered 5 different staterooms, sexually assaulting 5 women in the same night. Crewmember was caught by a victim's cabin mate and was severely beaten by cruise ship security while trying to jump ship. At the next port, Bermuda police came on board to collect evidence and statements. Trial was held in Bermuda as it was the governing jurisdiction and law applied. The court ruled that the assaulter was guilty on all 5 accounts and was sentenced to 9 months in jail for each assault, which he served concurrently before he was deported back to his home country. Of the five victims, three settled and two brought lawsuits against the cruise line for negligence. Forum-selection, notice-requirement, and choice-of-law clauses were all enforced in court. Because punitive damages were not available under the Jones Act (1983) or the Death on the High Seas Act (1983), the court reasoned that the policy of maintaining uniformity in maritime law dictated a finding that punitive damages were not available under federal general maritime law. Since allowing punitive damages under state law would have antagonized the policy of uniformity, and federal maritime law set the standards for liability in the passengers' personal injury actions, the court held that punitive damages were unavailable to the passengers. The jury found the cruise line was not liable at all to one of the victims, and liable to the other in an amount ($20,000). The court found that the cruise line was not negligent at all to one of the victims because she was the first of the five assaulted, thus the cruise line insinuated there was never a breach of contract. The other victim was the last of the five victimized and thus there was a breach in the duty of care owed. She was granted $20,000 for her past pain and suffering and nothing for her future pain and suffering.

All victims interviewed were asked at some point to sign a gag order and refused. These women are rare, most victims of sexual assault settle with the cruise line, and per the gag order commit to never talking about the assault or the recompense.

Conclusions

"A not uncommon problem is the allegation of sexual assault on passengers by crewmembers, particularly cabin, table or bar stewards. Members must have rigorous policies prohibiting socializing between crew and passengers. Any crewmember, found in a passenger area where he should not be, should be subject to dismissal for the first offense. A cruise operator must take reasonable care to investigate a crewmember's background before hiring him. There is at present a split of authority between US courts as to whether cruise operators are strictly liable for assaults by crew or whether operators are only liable if they are negligent in hiring or supervising crew members. The majority of the circuits have decided that operators are only liable for negligence. (A Guide to P&I Cover, The Standard, 2007, Testimony by Ross Klein on June 19th, 2008).

Court case outcomes would have been different had assaults been properly investigated and U.S. laws applied. Failure to collect evidence and the improper handling of evidence is a significant problem in regards to cruise ship crime. If authority personnel onboard are properly trained and investigations completed in a timely and thorough manner, victims would have a better chance of conviction. Cabins are routinely cleaned twice a day, using products that are chlorine bleach based to prevent outbreaks of Noroviruses. The use of such products destroys pertinent evidence during the delay between attack and landing at port.

Cruise lines need to have risk management procedures for handling sexual assault cases. Instead of firing a crewmember and sending them to their home country, they need to be questioned, trialed, and convicted. Paper trails do not exist without conviction, making it easy for crewmembers to gain employment under a different cruise line immediately following termination. Serial rapists should not have the ability to move from company to company and habitually assault women. In one case, for example, "it was subsequently learned that the crewmember had previously worked for a different cruise line that had 'do not rehire' marked on his personnel file. But he passed background checks and was hired by the current cruise line. It seems the cruise line also failed to notice that the name under which the man had applied for employment was different than the name on his passport" (Ross Klein, 2008).

"My greatest fear is that sexual offenders figure out that they can continuously go on cruise ships and take advantage of the lax regulations."
–Anonymous, 2009

Passengers who have been victims of cruise lines, whether it be assault, fraud, or deceptive trade practices, are unaware that their rights have been limited through the ticket-passenger contract.

It is disheartening that individuals interviewed found out they unknowingly entered a contract forfeiting and limiting their rights. "He (the attorney) had us bring the ticket book (victim's friend's name and I) and he opened it to the back, and he's the one who read to us that for any theft... or anything really, that happens while out to sea, would have to be trialed in Florida, as a civil or criminal case" (Anonymous, 2009). It is "fundamentally unfair to the unsuspecting passenger who does not realize the clauses exist until the ticket has already been purchased".

"In very small print, but yea... it's in there."

-Anonymous, 2009

"Currently, U.S. federal law enforcement agencies, including the FBI, Navy, and Coast Guard, must obtain permission from a cruise ship's flag country before beginning a criminal investigation; U.S. agents cannot even board a cruise ship that may be the scene of a crime committed by or against an American citizen if the ship is located beyond U.S. waters".

The Cruise Line International Association claims that cruise ships are inherently secure because ships offer a controlled environment with limited access. "However, there has been some startling statistics between 2003 and 2005: 24 people were reported missing and 178 people reported a claim against sexual assault. Additionally, the FBI has opened investigations on 305 cruise-based crimes, from 2000-2005". The CLIA compares these statistics to U.S. crime rates and harps on being the safest form of transportation and inherently secure. They fail however, to examine the context to which these statistics apply.

Ross Klein, a professor at Memorial University of Newfoundland, conducted a study based on the total number of passengers, total number of passenger nights, and the likelihood that they will be sexually assaulted during that time. Klein testified at a congressional hearing on March 27th, 2007.

During his testimony he stated:

"I conclude that one has a 50% greater chance of sexual assault on a Royal Caribbean International ship as compared to the US generally and that the pattern on Royal Caribbean International is consistent with that of other mass market cruise lines operating ships of similar size and style... a rate that is 48.065 sexual assaults per 100,000... The U.S. rate of sexual assaults... is 32.2 per 100,000."

Future Progress

On the bright side, cruise ship crime is finally starting to receive media attention and is slowly beginning to create awareness. The highly publicized disappearance of 26-year-old George Allen Smith IV on his honeymoon aboard the Brilliance of the Sea in July of 2005 was the final straw that pushed cruise related crime to the forefront. Jennifer, (George's wife) settled with the cruise line the same day his parents brought a lawsuit against the cruise line for deliberately portraying the incident as an accident, intentionally destroying evidence, and failing to not only properly investigate the crime, but failure to accurately report details to the FBI.

Congressman Christopher Shays not only pushed but also chaired the first two congressional hearings regarding cruise ship crime. On March 27th, 2007 during the congressional hearing "Crimes Against Americans on Cruise Ships" Ross Klein's testimony explained the high propensity for sexual assault aboard cruise ships. Fortunately, progress was made in April 2007 when the industry voluntarily signed an agreement with the FBI and the Coast Guard to report all incidents against Americans. "The only apparent weakness of this approach is that it does not define what constitutes a crime. This is important given the industry's propensity for manipulating definitions" (Ross Klein, 2008). Furthermore, because the agreement was a voluntary compliance, cruise lines cannot be penalized for failure to report crimes. Taking into account cruise lines were already required to report crimes against U.S. citizens prior to the agreement; it is unlikely this voluntary compliance will motivate them to radically change their behaviours and accurately portray cruise ship crime.

Although the industry has been firmly resistant to change and continuously denies that a problem exists, their actions acknowledge they are well aware. Royal Caribbean Cruise Line (RCCL) recently installed peepholes in all passenger stateroom doors of their entire fleet. Whether Royal Caribbean's action was motivated by pressure from an increasing public awareness or their personal desire to keep passengers safe is unknown. However, it is important to note that RCCL did not install deadbolts or top locks along with the peepholes. Thus, crewmembers possessing keycards are no less prohibited from making an attack than they were before. The industry needs to implement significant change in order to insure the safety of its passengers.

Presently, Representative Doris Matsui of the 5th district of California introduced a bill, The Cruise Vessel Security and Safety Act of 2009 on March 12, 2009 (S. 588, 2009). This bill has been introduced, in some form every year since 2004, and has never been voted on. The bill was prepared in conjunction with suggestions made by the International Cruise Victims (ICV) association during previous congressional hearings. Areas addressed in the bill are: improving ship safety, providing transparency in reporting, improving crime scene response, improving training procedures, enforcing safety and environmental standards, and amending the Death on the High Seas Act (1983). The amendments of this bill are of the utmost importance and need to be passed.

Even if passed, the cruise industry has a long way to go to insuring safety at a high degree of care. Passengers at sea need to be as equally protected as they are on land. U.S. citizens should not be allowed to unknowingly enter contracts forfeiting the same rights and remedies customary to common law.

Moving in the Right Direction

Only recently has the court mandated that cruise lines adhere to the guidelines established by the Americans with Disabilities Act (1990). The U.S. Supreme Court ruling was based on a class action lawsuit pursuant to Title III of the ADA. It was decided that foreign flagged vessels must comply with the public accommodations requirements established by the ADA. For the most part, the U.S. Supreme Court adheres to federal maritime law, removing itself from interference. It is doubtful this law would have passed had it not been such a severe case of injustice against civil rights.

Hopefully, through awareness and persistence, the cruise industry will feel the pressure to change for the benefit of everyone. Cruise lines have not been successful in preventing devastating experiences. Taking a cruise can be the experience and memory of a lifetime. It is imperative that this is the case for everyone, and to not brush aside those who have suffered. The interviewed victims were individuals taking a cruise for the same reason as everyone, to have a fun and enjoyable experience. Instead, they suffered an unthinkable tragedy. It is clearly evident that if it happened to those interviewed it could happen to anyone.

"Passengers check their brain at the door, thinking it's a perfectly safe environment, but it's not..."

2

Security from Somalian Sea-robbers

Piracy in Somalia

Piracy off the coast of Somalia has been a threat to international shipping since the second phase of the Somali Civil War in the early 21st century. Since 2005, many international organizations, including the International Maritime Organization and the World Food Programme, have expressed concern over the rise in acts of piracy. Piracy has impeded the delivery of shipments and increased shipping expenses, costing an estimated £10 billion a year in global trade. According to the German Institute for Economic Research (DIW), a veritable industry of profiteers has also risen around the piracy. Insurance companies, in particular, have profited from the pirate attacks, as insurance premiums have increased exponentially.

A United Nations report and several news sources have suggested that piracy off the coast of Somalia is caused in part by illegal fishing. According to the DIW and the U.S. House Armed Services Committee, the dumping of toxic waste in Somali waters by foreign vessels has also severely constrained the ability of local fishermen to earn a living and forced many to turn to piracy instead. Other articles allege that 70 percent of the local coastal communities "strongly support the piracy as a form of national defence of the country's territorial waters", and that the pirates believe they are protecting their fishing grounds and exacting justice and compensation for the marine resources stolen. Some pirates have suggested that, in the absence of an effective national coast guard following the outbreak of the Somali Civil War and the subsequent disintegration of the Armed Forces, they became pirates in order to protect their waters. This belief is also reflected in the names taken on by some of the pirate networks, such as the *National Volunteer Coast Guard* (NVCG). However, as piracy has become substantially more lucrative in recent years, some reports are suggesting that financial gain is now the primary motive for Somali pirates.

Combined Task Force 150, a multinational coalition task force, took on the role of fighting Somali piracy by establishing a Maritime Security Patrol Area (MSPA) within the Gulf of Aden. The increasing threat posed by piracy has also caused concern in India since most of its shipping trade routes pass through the Gulf of Aden. The Indian Navy responded to these concerns by deploying a warship in the region on 23 October 2008. In September 2008, Russia announced that it too would join international efforts to combat piracy. Some reports have also accused certain government officials in Somalia of complicity with the pirates, with authorities from the Galmudug administration in the north-central Hobyo district reportedly attempting to use pirate gangs as a bulwark against Islamist insurgents from the nation's southern conflict zones. However, according to UN Secretary-General Ban Ki Moon, both the former and current administrations of the autonomous Puntland region in northeastern Somalia appear to be more actively involved in combating piracy. The latter measures include on-land raids on pirate hideouts, and the construction of a new naval base in conjunction with Saracen International, a UK-based security company. By the first half of 2010, these increased policing efforts by Somali government authorities on land and international naval vessels at sea reportedly contributed to a drop in pirate attacks in the Gulf of Aden from 86 a year prior to 33, forcing pirates to shift attention to other areas such as the Somali Basin and the wider Indian Ocean. As of 11 December 2010, Somali pirates were holding at least 35 ships with more than 650 hostages.

History

During the Siad Barre regime, Somalia received aid from Denmark, Great Britain, Iraq, Japan, Sweden, USSR and West

Germany to develop its fishing industry. Cooperatives had fixed prices for their catch, which was often exported due to the low demand for seafood in Somalia. Aid money improved the ships and supported the construction of maintenance facilities. After the fall of the Barre regime, the income from fishing decreased due to the Somali Civil War.

Also, there was no coast guard to protect against fishing trawlers from other countries illegally fishing and big companies dumping waste which killed fish in Somali waters. This led to the erosion of the fish stock. Local fishermen started to band together to protect their resources. Due to the clan-based nature of Somali society, the lack of a central government and Somalia's strategic location at the Horn of Africa, conditions were ripe for the growth of piracy in the early 1990s.

Precise data on the current economic situation in Somalia is scarce but with an estimated per capita GDP of $600 per year, it remains one of the world's poorest countries. Millions of Somalis depend on food aid and in 2008, according to the World Bank, as much as 73% of the population lived on a daily income below $2. These factors and the lucrative success of many hijacking operations have drawn a number of young men toward gangs of pirates, whose wealth and strength often make them part of the local social and economic elite. Abdi Farah Juha who lives in Garoowe (100 miles from the sea) told the BBC, "They have money; they have power and they are getting stronger by the day. They wed the most beautiful girls; they are building big houses; they have new cars; new guns."

Some pirates are former fishermen, whose livelihoods were hurt by foreign ships illegally fishing in Somali waters. Most of the pirates, observers say, are not former fishermen. After seeing the profitability of piracy, since ransoms are usually paid, warlords began to facilitate pirate activities, splitting the profits with the pirates. Pirates even attack ships carrying humanitarian aid. In most of the hijackings, the bandits have not harmed their prisoners.

The Transitional Federal Government has made some efforts to combat piracy, occasionally allowing foreign naval vessels into Somali territorial waters. However, more often than not, foreign naval vessels chasing pirates were forced to break off when the pirates entered Somali territorial waters. To counter this, in 2008 (and renewed each year since then) the UN passed a resolution allowing international warships to pursue pirates into Somali territorial waters. On the advice of lawyers, the Royal Navy and other international naval forces have often released suspected pirates that they have captured because, although the men are frequently armed, they have not been caught engaging in acts of piracy and have thus not technically committed a crime. The government of Puntland has made more progress in combating piracy, evident in recent interventions.

Summary of Recent Events

Somali pirates have attacked hundreds of vessels in the Arabian Sea and Indian Ocean region, though most attacks do not result in a successful hijacking. In 2008, there were 111 attacks which included 42 successful hijackings. However, this is only a fraction the up to 30,000 merchant vessels which pass through that area. The rate of attacks in January and February 2009 was about 10 times higher than during the same period in 2008 and "there have been almost daily attacks in March", with 79 attacks, 21 successful, by mid April. Most of these attacks occur in the Gulf of Aden but the Somali pirates have been increasing their range and have started attacking ships as far south as off the coast of Kenya in the Indian Ocean. Below are some notable pirate events which have garnered significant media coverage since 2007.

On 28 May 2007, A Chinese sailor was killed by the pirates because the ship's owners failed to meet their ransom demand.

On 5 October 2008, the United Nations Security Council adopted resolution 1838 calling on nations with vessels in the area to apply military force to repress the acts of piracy. At the 101st council of the International Maritime Organization, India called for a United Nations peacekeeping force under unified command to tackle piracy off Somalia. (There has been a general and complete arms embargo against Somalia since 1992.)

On 21 November 2008, BBC News reported that the Indian Navy had received United Nations approval to enter Somali waters to combat piracy. In November 2008, Somali pirates began hijacking ships well outside the Gulf of Aden, perhaps targeting ships headed for the port of Mombasa, Kenya. The frequency and sophistication of the attacks also increased around this time, as did the size of vessels being targeted. Large cargo ships, oil and chemical tankers on international voyages became the new targets of choice for the Somali hijackers.

This is in stark contrast to the pirate attacks which were once frequent in the Strait of Malacca, another strategically important waterway for international trade, which were according to maritime security expert Catherine Zara Raymond, generally directed against "smaller, more vulnerable vessels carrying trade across the Straits or employed in the coastal trade on either side of the Straits."

On 19 November 2008, the Indian Navy warship INS Tabar sank a suspected pirate mothership. Later, it was claimed to be a Thai trawler being hijacked by pirates. The Indian Navy later defended its actions by stating that they were fired upon first.

On 8 April 2009, four Somali pirates seized the *Maersk Alabama* 240 nautical miles (440 km; 280 mi) southeast of the Somalia port city of Eyl. The ship was carrying 17,000 metric tons of cargo, of which 5,000 metric tons were relief supplies bound for Somalia, Uganda, and Kenya.

On 12 April 2009, United States Navy SEAL snipers killed the three pirates that were holding Captain Richard Phillips hostage aboard a lifeboat from the *Maersk Alabama* after determining that Captain Phillips' life was in immediate danger. A fourth pirate, Abdul Wali Muse, surrendered and was taken into custody. On May 18, a federal grand jury in New York returned a ten-count indictment against him.

On 20 April 2009, United States Secretary of State Hillary Clinton commented on the capture and release of 7 Somali pirates by Dutch Naval forces who were on a NATO mission. After an attack on the *Handytankers Magic,* a petroleum tanker, the Dutch frigate *De Zeven Provinciën* tracked the pirates back to a pirate "mother ship" and captured them. They confiscated the pirates weapons and freed 20 Yemeni fishermen whom the pirates had kidnapped and who had been forced to sail the pirate "mother ship". Since the Dutch Naval Forces were part of a NATO exercise, but not on an EU mission, they lacked legal jurisdiction to keep the pirates so they released them. Clinton stated that this action "sends the wrong signal" and that additional coordination was needed among nations.

On 23 April 2009, international donors pledged over $250 million for Somalia, including $134 million to increase the African Union peacekeeping mission from 4,350 troops to 8,000 troops and $34 million for Somali security forces. Secretary-General of the United Nations Ban Ki-moon told delegates at a donors' conference sponsored by the U.N. that "Piracy is a symptom of anarchy and insecurity on the ground", and that "More security on the ground will make less piracy on the seas." Somali President Sharif Ahmed pledged at the conference that he would fight piracy and to loud applause said that "It is our duty to pursue these criminals not only on the high seas, but also on terra firma". The Somali government has not gone after pirates because pirate leaders currently have more power than the government. It has been estimated by piracy experts that in 2008 the pirates gained about $80 million through ransom payments.

On 2 May 2009, Somali pirates captured the MV *Ariana* with its 24 Ukrainian crew. The ship was released on 10 December 2009 after a ransom of almost $3,000,000 was paid.

On 8 November 2009, Somali pirates threatened that a kidnapped British couple, the Chandlers, would be "punished" if a German warship did not release seven pirates. Omer, one of the pirates holding the British couple, claimed the seven men were fishermen, but a European Union Naval Force spokesman stated they were captured as they fired AK-47 assault rifles at a French fishing vessel. The Chandlers were released on 14 November 2010 after 388 days of captivity. At least two ransom payments, reportedly over GBP 500 000, had been made.

April 2010, the Central Intelligence Agency (CIA) alluded to possible covert and overt action against the pirates. CIA officials had been publicly warning of this potential threat for months. In a *Harpers Magazine* article, a CIA official said, "We need to deal with this problem from the beach side, in concert with the ocean side, but we don't have an embassy in Somalia and limited, ineffective intelligence operations. We need to work in Somalia and in Lebanon, where a lot of the ransom money has changed hands. But our operations in Lebanon are a joke, and we have no presence at all in Somalia."

On 11 May 2010, Somali pirates seized a Bulgarian-flagged ship in the Gulf of Aden. The *Panega,* with 15 Bulgarian crew members aboard, was en route from the Red Sea to India or Pakistan. This was the first such hijacking of a Bulgarian-flagged ship. On 12 May 2010, Athens announced that Somali pirates have seized a Greek vessel in the Gulf of Aden with at least 24 people onboard, including two Greek citizens and some Filipinos. The vessel, sailing under the Liberian flag, was transporting iron from Ukraine to China.

On 14 January 2011, while speaking to reporters, Commodore Michiel Hijmans of the Royal Netherlands Navy stated that the use of hijacked vessels in more recent hijackings had led to increased range of pirating activities, as well as difficulty to actively thwart future events due to the use of kidnapped sailors as human shields.

On 15 January 2011, thirteen Somali pirates seized the *Samho Jewelry* a Maltese-flagged chemical carrier operated by Samho Shipping 650 km southeast of Muscat. The Republic of Korea Navy destroyer *Choi Young* shadowed the *Samho Jewlry* for several days. In the early morning of 21 January 2011, 25 ROK Navy SEALs on small boats launched from the *Choi Young* boarded the *Samho Jewelry* while the *Choi Young*s Westland Super Lynx provided covering fire. Eight pirates were killed and five captured in the operation, the crew of 21 was freed with the Captain suffering a gunshot wound to the stomach.

On 28 January 2011, an Indian Coast Guard aircraft while responding to a distress call from the CMA CGM *Verdi,* located two skiffs attempting a piracy attack near Lakshadweep. Seeing the aircraft, the skiffs immediately aborted their piracy attempt and dashed towards the mother vessel, MV *Prantalay 14*-a hijacked Thai trawler, which hurriedly hoisted the two skiffs onboard and moved westward. The Indian Navy deployed the INS *Cankarso* which located and engaged the mothership 100 nautical miles north of the Minicoy island. 10 pirates where killed while 15 were apprehended and 20 Thai and Myanmarese fishermen being held aboard the ship as hostages were rescued. Within a week of its previous success, the Indian Navy captured another hijacked Thai trawler, MV *Prantalay 11* and captured 28 pirates aboard in an operation undertaken by the INS *Tir* purusuant to receiving information that a Greek merchant ship had been attacked by pirates on board high-speed boats, although it had managed to avoid capture. When INS *Tir* ordered the pirate ship to stop and be boarded for inspection, it was fired upon. The INS *Tir* returned fire in which 3 pirates were injured and caused the pirates to raise a white flag indicating their surrender. The INS *Tir* subsequently joined by CGS *Samar* of the Indian Coast Guard. Officials from the Indian Navy reported that a total of 52 men were apprehended, but that 24 are suspected to be Thai fishermen who were hostages of the 28 African pirates.

In late February 2011, piracy targeting smaller yachts and collecting ransom made headlines when four Americans were killed aboard their vessel by their captors, while a military ship shadowed them. On 24 February 2011 a Danish family on a yacht were captured by pirates. In March 2011, the Indian Navy intercepted a pirate mother vessel 600 nautical miles west of the Indian coast in the Arabian Sea on Monday and rescued 13 hostages. Sixty-one pirates have also been caught in the operation carried out by Navy's INS *Kalpeni*. In late March 2011, Indian Navy seized 16 Somali pirates after a three-hour-long battle in the Arabian Sea, The navy also rescued 16 crew members of a hijacked Iranian ship west of the Lakshadweep Islands. The crew included 12 Iranians and four Pakistanis.

In early May 2011, Russian special forces retook a Russian oil tanker that had been hijacked by 11 pirates. One died in the assault, and a week later Russian military official reported that the remainder were freed due to weaknesses in international law but died before reaching the Somali coast. Russian President Dmitry Medvedev had announced the day the ship was retaken that "We'll have to do what our forefathers did when they met the pirates" until a suitable way of prosecuting them was available.

Pirates

Profile

Many pirates are 20–35 years old and come from the region of Puntland, in northeastern Somalia. The East African Seafarers'

Association estimates that there are at least five pirate gangs and a total of 1,000 armed men. According to a BBC report, the pirates can be divided into three main categories:

- Local Somali fishermen, considered the brains of the pirates' operations due to their skill and knowledge of the sea. Many think that foreign boats have no right to cruise next to the shore and destroy their boats.
- Ex-militiamen, who previously fought for the local clan warlords, or ex-military from the former Barre government used as the muscle.
- Technical experts, who operate equipment such as GPS devices.

According to Globalsecurity.org, there are four main groups operating off the Somali coast. The National Volunteer Coast Guard, commanded by Garaad Mohamed, is said to specialize in intercepting small boats and fishing vessels around Kismayo on the southern coast. The Marka group, under the command of Yusuf Mohammed Siad Inda'ade, is made up of several scattered and less organized groups operating around the town of Marka. The third significant pirate group is composed of traditional Somali fishermen operating around Puntland and referred to as the Puntland Group. The last set are the Somali Marines, reputed to be the most powerful and sophisticated of the pirate groups with a military structure, a fleet admiral, admiral, vice-admiral and a head of financial operations.

The conduct of a typical pirate attack has been analyzed and shows that while attacks can be expected at any time, most occur during the day, often in the early hours. They may involve two or more skiffs that can reach speeds of up to 25 knots. With the help of motherships that include captured fishing and merchant vessels the operating range of the skiffs has been increased far into the Indian Ocean. An attacked vessel is approached from quarter or stern; RPGs and small arms are used to intimidate the operator to slow down and allow boarding. Light ladders are brought along to climb aboard. Pirates then will try and get control of the bridge to take operational control of the vessel. The closest Somali term for 'pirate' is *burcad badeed,* which means "ocean robber". But the pirates themselves prefer to be called *badaadinta badah,* or "saviours of the sea", or in the English "coastguard", as they claimed that their action was for "protection of his sea, the native [Somali] waters".

Effects and Perceptions

There have been both positive and negative effects of the pirates' economic success. Local residents have complained that the presence of so many armed men makes them feel insecure, and that their free spending ways cause wild fluctuations in the local exchange rate. Others fault them for excessive consumption of alcoholic beverages and khat.

On the other hand, many other residents appreciate the rejuvenating effect that the pirates' on-shore spending and re-stocking has had on their impoverished towns, a presence which has often provided jobs and opportunity when there were none. Entire hamlets have in the process been transformed into veritable boomtowns, with local shop owners and other residents using their gains to purchase items such as generators — "allowing full days of electricity, once an unimaginable luxury."

Local fishermen in the Malindi area of Kenya to the south have reported their largest catches in forty years, catching hundreds of kilos of fish and earning fifty times the average daily wage as a result. They attribute the recent abundance of marine stock to the pirates scaring away the foreign fishing trawlers, which it is claimed have for decades deprived local dhows of a livelihood. Marine biologists agree, saying that the indicators are that the local fishery is recovering because of the lack of commercial scale fishing.

The Somalian piracy appears to have a positive impact on the problem of overfishing in Somali waters by foreign vessels, as a comparison has been made with the situation in Tanzania further to the south, which suffers from the same problem, and also lacks the means to enforce the protection and regulation of its territorial waters. There, the catches have dropped to dramatic low levels, whereas in Somalia they have risen back to more acceptable levels since the beginning of the piracy.

Weaponry and Funding

The pirates get most of their weapons from Yemen, but a

significant amount come from Mogadishu, Somalia's capital. Weapons dealers in the capital receive a deposit from a *hawala* dealer on behalf of the pirates and the weapons are then driven to Puntland where the pirates pay the balance. Various photographs of pirates in situ indicate that their weapons are predominantly AKMs, RPG-7s, AK47s, and semi-automatic pistols such as the TT-30. Additionally, given the particular origin of their weaponry, they are likely to have hand grenades such as the RGD-5 or F1. Al-Qaeda reportedly funded pirates with cash to purchase weapons. Osama bin-Laden supported these pirates in a video footage aired on al-Jazeera.

The funding of piracy operations is now structured in a stock exchange, with investors buying and selling shares in upcoming attacks in a bourse in Harardhere. Pirates say ransom money is paid in large denomination US dollar bills. It is delivered to them in burlap sacks which are either dropped from helicopters or cased in waterproof suitcases loaded onto tiny skiffs. Ransom money has also been delivered to pirates via parachute, as happened in January 2009 when an orange container with $3 million cash inside was dropped onto the deck of the supertanker MV *Sirius Star* to secure the release of ship and crew. To authenticate the banknotes, pirates use currency-counting machines, the same technology used at foreign exchange bureaus worldwide. According to one pirate, these machines are, in turn, purchased from business connections in Dubai, Djibouti, and other areas. Hostages seized by the pirates usually have to wait 45 days or more for the ships' owners to pay the ransom and secure their release.

Somali pirates allegedly get help from the Somali diaspora. Somali expatriates, including reputedly some among the 200,000 Somalis living in Canada, offer funds, equipment and information.

Economic Issues

General

The purpose of piracy is to get ransom money for release of the crew, ship, and cargo. Pirates' income from ransom has been estimated to be about 39 million euro (about $58 million) in 2009 and $238 million in 2010. However, indirect costs of piracy are much higher and estimated to be between $7 to 12 billion as they also include insurance, naval support, legal proceedings, re-routing of slower ships, and individual protective steps taken by ship-owners. Further, piracy in Somalia leads to a decrease of revenue for Egypt as fewer ships use the Suez canal (estimated loss of about $642 million), impedes trade with a number of countries such as Kenya and Yemen, and is detrimental to tourism and fishing in the Seychelles.

A 2011 report published by Geopolicity Inc, investigated the causes and consequences of international piracy, with a particular focus on piracy emanating from Somalia. The report asserts that piracy is an emerging market in its own right, valued at between US$4.9-8.3 billion in 2010 alone, and it establishes, for the first time, an economic model for assessing the costs and benefits of international piracy. This model provides a comprehensive, independent framework of trend analysis, whilst also highlighting where the greatest rates of return on international counter pirate investment and policy are to be found across what Geopolicity term the 'Pirate Value Chain.' The report states that the number of pirates could double by 2016, increasing by 400 each year. This is being fuelled by attractive financial incentives with Somali pirates earning up to US$79,000/year; equating to almost 150 times their country's national average wage.

Human Cost

Somali piracy operations extol significant human cost. Thus in 2010, 4,185 seafarers had been attacked and 1,090 were held hostage, a third of these were abused. According to Reuters, 62 seafarers died out of over 3,500 captives during a four year period having been murdered, or from suicide or malnutrition. Many seafarers are traumatized after release.

On the Somali side, youth is directed into a perilous life of criminal activity.

Profiteers

According to the German Institute for Economic Research (DIW), a veritable industry of profiteers has also risen around the piracy. Insurance companies, in particular, have profited from the pirate attacks, as insurance premiums have increased exponentially.

To facilitate the continuation of such monetary gains, insurance firms have actively discouraged ship owners from taking security precautions. For their part, shipping companies have often rebuffed naval guidelines on how best to prevent pirate attacks so as to cut down on costs. Ship crews have also been reluctant to repel the pirates on account of their low wages and inequitable work contracts. In addition, security contractors, including the German arms industry, have profited from the phenomenon.

Sovereignty and Environmental Protection

The former UN envoy for Somalia, Ahmedou Ould-Abdallah, has stated that "because there is no (effective) government, there is... much irregular fishing from European and Asian countries," and that the UN has what he described as "reliable information" that European and Asian companies are dumping toxic and nuclear waste off the Somali coastline. However, he stresses that "no government has endorsed this act, and that private companies and individuals acting alone are responsible." In addition, Ould-Abdallah told the press that he approached several international NGOs, such as Global Witness, to trace the illicit fishing and waste-dumping. He added that he believes the toxic waste dumping is "a disaster off the Somali coast, a disaster (for) the Somali environment, the Somali population", and that what he terms "this illegal fishing, illegal dumping of waste" helps fuel the civil war in Somalia since the illegal foreign fishermen pay off corrupt local officials or warlords for protection or to secure counterfeit licenses. Ould-Abdallah noted that piracy will not prevent waste dumping:

> I am convinced there is dumping of solid waste, chemicals and probably nuclear (waste).... There is no government (control) and there are few people with high moral ground[...] The intentions of these pirates are not concerned with protecting their environment. What is ultimately needed is a functioning, effective government that will get its act together and take control of its affairs.
>
> —Ahmedou Ould-Abdallah, the UN envoy for Somalia

Somali pirates which captured MV *Faina*, a Ukrainian ship carrying tanks and military hardware, accused European firms of dumping toxic waste off the Somali coast and declared that the $8m ransom for the return of the ship will go towards cleaning up the waste. The ransom demand is a means of "reacting to the toxic waste that has been continually dumped on the shores of our country for nearly 20 years", Januna Ali Jama, a spokesman for the pirates said. "The Somali coastline has been destroyed, and we believe this money is nothing compared to the devastation that we have seen on the seas."

These issues have generally not been reported in international media when reporting on piracy. According to Muammar al-Gaddafi, "It is a response to greedy Western nations, who invade and exploit Somalia's water resources illegally. It is not a piracy, it is self defence."

Pirate leader Sugule Ali said their motive was "to stop illegal fishing and dumping in our waters... We don't consider ourselves sea bandits. We consider sea bandits [to be] those who illegally fish and dump in our seas and dump waste in our seas and carry weapons in our seas." Also, the independent Somalian news-site WardherNews found that 70 percent "strongly supported the piracy as a form of national defence of the country's territorial waters".

Waste Dumping

Following the Indian Ocean tsunami of December 2004, there have emerged allegations that after the outbreak of the Somali Civil War in late 1991, Somalia's long, remote shoreline was used as a dump site for the disposal of toxic waste. The huge waves which battered northern Somalia after the tsunami are believed to have stirred up tonnes of nuclear and toxic waste that was illegally dumped in Somali waters by several European firms — front companies created by the Italian mafia. The European Green Party followed up these revelations by presenting before the press and the European Parliament in Strasbourg copies of contracts signed by two European companies—the Italian Swiss firm, Achair Partners, and an Italian waste broker, Progresso—and representatives of the warlords then in power, to accept 10 million tonnes of toxic waste in exchange for $80 million (then about £60 million). According to a report by the United Nations Environment Programme (UNEP) assessment mission, there are far higher than normal cases of respiratory infections, mouth ulcers and bleeding, abdominal hemorrhages and unusual skin infections among many inhabitants of the areas around the northeastern towns of Hobbio and Benadir on the Indian Ocean coast—diseases consistent with radiation sickness. UNEP continues that the current situation along the Somali coastline poses a very serious environmental hazard not only in Somalia but also in the eastern Africa sub-region.

In 1992, reports ran in the European press of "unnamed European firms" contracting with local warlords to dump toxic waste both in Somalia and off Somalia's shores. The United Nations Environment Program was called in to investigate, and the Italian parliament issued a report later in the decade. Several European "firms" — really front companies created by the Italian mafia — contracted with local Somali warlords to ship hundreds of thousands of tons of toxic industrial waste from Europe to Somalia.

—Troy S. Thomas, *Warlords rising: confronting violent non-state actors*

Under Article 9(1)(d) of the Basel Convention on the Control of Transboundary Movements of Hazardous Wastes and Their Disposal, it is illegal for "any transboundary movement of hazardous wastes or other wastes: that results in deliberate disposal (e.g. dumping) of hazardous wastes or other wastes in contravention of this Convention and of general principles of international law".

According to Nick Nuttall of the United Nations Environmental Programme, "Somalia has been used as a dumping ground for hazardous waste starting in the early 1990s, and continuing through the civil war there", and "European companies found it to be very cheap to get rid of the waste, costing as little as $2.50 a tonne, where waste disposal costs in Europe are something like $1000 a tonne."

Illegal Fishing

At the same time, foreign travellers began illegally fishing Somalia's seas, with an estimated $300 million of tuna, shrimp, and lobster being taken each year, depleting stocks previously available to local fishermen. Through interception with speedboats, Somali fishermen tried to either dissuade the dumpers and trawlers or levy a "tax" on them as compensation, as Segule Ali's previously mentioned quote notes. Peter Lehr, a Somalia piracy expert at the University of St. Andrews says *"It's almost like a resource swap, Somalis collect up to $100 million a year from pirate ransoms off their coasts and the Europeans and Asians poach around $300 million a year in fish from Somali waters."* The UK's Department for International Development (DFID) issued a report in 2005 stating that, between 2003 and 2004, Somalia lost about $100 million in revenue due to illegal tuna and shrimp fishing in the country's exclusive economic zone by foreign trawlers.

According to Roger Middleton of Chatham House, "The problem of overfishing and illegal fishing in Somali waters is a very serious one, and does affect the livelihoods of people inside Somalia [...] the dumping of toxic waste on Somalia's shores is a very serious issue, which will continue to affect people in Somalia long after the war has ended, and piracy is resolved." To lure fish to their traps, foreign trawlers reportedly also use fishing equipment under prohibition such as nets with very small mesh sizes and sophisticated underwater lighting systems.

Under Article 56(1)(b)(iii) of the Law of the Sea Convention:

> *"In the exclusive economic zone, the coastal State has jurisdiction as provided for in the relevant provisions of this Convention with regard to the protection and preservation of the marine environment".*

Article 57 of the Convention in turn outlines the limit of that jurisdiction:

> *"The exclusive economic zone shall not extend beyond 200 nautical miles from the baselines from which the breadth of the territorial sea is measured".*

Anti-piracy Measures

Self-defence

The third volume of the handbook: *Best Management Practices to Deter Piracy off the Coast of Somalia and in the Arabian Sea Area* (known as BMP3) is the current authoritative guide for merchant ships on self-defence against pirates. The guide is issued and updated by a consortium of interested international shipping and trading organizations including the EU, NATO and the

International Maritime Bureau. It is distributed primarily by the Maritime Security Centre – Horn of Africa (MSCHOA)-the planning and coordination authority for EU naval forces (EUNAVFOR). BMP3 encourages vessels to register their voyages through the region with MSCHOA as this registration is a key component of the operation of the International Recommended Transit Corridor (IRTC) (the navy patrolled route through the Gulf of Aden). BMP3 also contains a chapter entitled "Self-Protective Measures" which lays out a list of steps a merchant vessel can take on its own to make itself less of a target to pirates and make it better able to repel an attack if one occurs. This list includes doing thing like ringing the deck of the ship with razor wire, rigging fire-hoses to spray sea-water over the side of the ship (to hinder boardings), having a distinctive pirate alarm, hardening the bridge against gunfire and creating a "citadel" where the crew can retreat in the event pirates get on board. Other unofficial self-defence measures that can be found on merchant vessels include the setting up of mannequins posing as armed guards or firing flares at the pirates.

Though it varies by country, generally peacetime law in the 20th and 21st centuries has not allowed merchant vessels to carry weapons. As a response to the rise in modern piracy, however, the U.S. Government changed its rules so that it is now possible for US flagged vessels to embark a team of armed private security guards. Other countries and organisations have similarly followed suit. This has given birth to a new breed of private security companies who provide training and protection for crew members and cargo and have proved effective in countering pirate attacks. The USCG leaves it to ship owners' discretion to determine if those guards will be armed.

Military Presence

The military response to pirate attacks has brought about a rare show of unity by countries that are either openly hostile to each other, or at least wary of cooperation, military or otherwise. Currently there are three international naval task forces in the region, with numerous national vessels and task forces entering and leaving the region, engaging in counter-piracy operations for various lengths of time. The three international task forces which comprise the bulk of counter-piracy operations are Combined

Task Force 150 (whose overarching mission is Operation Enduring Freedom), Combined Task Force 151 (which was set up in 2009 specifically to run counter-piracy operations) and the EU naval task force operating under Operation Atalanta. All counter-piracy operations are coordinated through a monthly planning conference called Shared Awareness and Deconfliction (SHADE). Originally having representatives only from NATO, the EU, and the Combined Maritime Forces (CMF) HQ in Bahrain, it now regularly attracts representatives from over 20 countries.

As part of the international effort, Europe plays a significant role in combating piracy of the Horn of Africa. The European Union under the Common Security and Defence Policy (CSDP) launched EU NAVFOR Somalia-Operation Atalanta (In support of Resolutions 1814 (2008), 1816 (2008), 1838 (2008) and 1846 (2008) of the United Nations Security Council (UNSC)). This operation is working to protect humanitarian aid and reduce the disruption to the shipping routes and the de-stabilising of the maritime environment in the region. To date, 26 countries have brought some kind of contribution to the operation. 13 EU Member States have provided an operational contribution to EU NAVFOR, either with ships, with maritime patrol and reconnaissance aircraft, or with Vessel Protection Detachment (VPD) team. This includes France, Spain, Germany, Greece, Sweden, Netherlands, Italy, Belgium, United Kingdom (also hosting the EU NAVFOR Operational headquarters), Portugal, Luxembourg, Malta and Estonia. 9 other EU Member States have participated in the effort providing military staff to work at the EU NAVFOR Operational Headquarters (Northwood Headquarters – UK) or onboard units. These are Cyprus, Romania, Bulgaria, Slovenia, Czech Republic, Hungary, Poland, Ireland and Finland. Finally, 4 non EU Member States, Norway (who has also provided an operational contribution with a warship regularly deploying), Croatia, Ukraine and Montenegro have so far also brought their contribution to EU NAVFOR. At any one time, the European force size fluctuates according to the monsoon seasons, which determine the level of piracy. It typically consists of 5 to 10 Surface Combatants (Naval ships), 1 to 2 Auxiliary ships and 2 to 4 Maritime Patrol and Reconnaissance Aircraft. Including land-based personnel, Operation Atalanta consists of a total of around 2,000 military personnel. EU NAVFOR operates in a zone comprising the south of the Red Sea, the Gulf of Aden and the western part of the Indian Ocean including the Seychelles, which represents an area of 2,000,000 square nautical miles.

Additionally, there are, and have been, several naval deployments by non-multinational task forces in the past. Some notable ones include:

On 29 May 2009, Australia pledged its support, re-directing Australian Warship, HMAS *Warramunga* from duties in the Persian Gulf to assist in the fighting of Piracy.

On 26 December 2008, China dispatched three warships (*Haikou* (171), *Wuhan* (169) and the supply ship *Weishanhu*) to the Gulf of Aden. A team of 16 Chinese Special Forces members from its Marine Corps armed with attack helicopters were on board. Subsequent to their initial deployment, China has maintained a three-ship flotilla of two warships and one supply ship in the Gulf of Aden by assigning ships from the South Sea Fleet and/or East Sea Fleet to the area on a three month basis.

In response to the increased activity of the INS *Tabar*, India sought to augment its naval force in the Gulf of Aden by deploying the larger INS *Mysore* to patrol the area. Somalia also added India to its list of states, including the U.S. and France, who are permitted to enter its territorial waters, extending up to 12 nautical miles (22 km; 14 mi) from the coastline, in an effort to check piracy. An Indian naval official confirmed receipt of a letter acceding to India's prerogative to check such piracy. "We had put up a request before the Somali government to play a greater role in suppressing piracy in the Gulf of Aden in view of the United Nations resolution. The TFG government gave its nod recently." India also expressed consideration to deploy up to four more warships in the region. On 14 March 2011, the Indian navy reportedly had seized 61 pirates and rescued 13 crew from the vessel, which had been used as a mother ship from where pirates launched attacks around the Indian Ocean. Meanwhile, a Bangladeshi ship hijacked by pirates last year was freed after a ransom was paid.

Norway announced on 27 February 2009, that it would send the frigate HNoMS *Fridtjof Nansen* to the coast of Somalia to fight piracy. Royal Norwegian Navy *Fridtjof Nansen* joined EU NAVFOR's international naval force in August.

Russia also chose to send more warships to combat piracy near Somalia following the announcement from the International Maritime Bureau terming the menace as having gone "out of control."

Due to their proximity to Somalia, the coast guard of Seychelles has become increasingly involved in counter-piracy in the region. On 30 March 2010, a Seychelles Coast Guard Trinkat-class patrol vessel rescued 27 hostages and sank two pirate vessels.

Other non-NATO and non-EU countries have, at one time or another, contributed to counter-piracy operations. Pakistan, South Korea, Japan, Thailand, Malaysia, Saudi Arabia and Iran have all sent ships to the region, sometimes joining with the existing CTFs, sometimes operating independently.

A maritime conference was also held in Mombasa to discuss the rising concern of regional piracy with a view to give regional and world governments recommendations to deal with the menace. The International Transport Workers Federation (ITWF) organised the regional African maritime unions' conference, the first of its kind in Africa. Godfrey Matata Onyango, executive secretary of the Northern Corridor Transit Coordination Authority said that "We cannot ignore to discuss the piracy menace because it poses a huge challenge to the maritime industry and if not controlled, it threats to chop off the regional internal trade. The cost of shipping will definitely rise as a result of the increased war insurance premium due to the high risk off the Gulf of Aden." In 2008 Pakistan offered the services of the Pakistan Navy to the United Nations in order to help combat the piracy in Somalia "provided a clear mandate was given."

Current Fleet of Vessels in Operation

Vessels, aircraft and personnel whose primary mission is to conduct anti-piracy activities come from different countries and are assigned to the following missions: Ocean Shield (NATO and partner states), Atalanta (EU and partner states), Combined Task Force 151, independent missions of Japan, India, Russia, PR China, South Korea, Iran, Malaysia. Additionally resources dedicated for the War on Terror missions of Combined Task Force 150 and Enduring Freedom – Horn of Africa also operate against the pirates.

Current States that Contribute Aircraft or Military Personnel

This list is incomplete; you can help by expanding it.

Former Contributions During Previous Rotation Periods

Brian Murphy (Associated Press) reported on 8 January 2009 that Rear Admiral Terence E. McKnight, U.S. Navy, is to command a new multi-national naval force to confront piracy off the coast of Somalia. This new anti-piracy force was designated Combined Task Force 151 (CTF-151), a multinational task force of the Combined Maritime Forces (CMF). The USS *San Antonio* was designated as the flagship of Combined Task Force 151, serving as an afloat forward staging base (AFSB) for the following force elements:

- 14-member U.S. Navy visit, board, search and seizure (VBSS) team.
- 8-member U.S. Coast Guard Law Enforcement Detachment (LEDET) 405.
- Scout Sniper Platoon, 2nd Battalion, 6th Marine Regiment, 26th Marine Expeditionary Unit (26 MEU) cross-decked from the USS *Iwo Jima* (LHD-7).
- 3rd platoon of the 2nd Battalion, 6th Marine's 'Golf' Infantry Company, a military police detachment, and intelligence personnel.
- Fleet Surgical Team 8 with level-two surgical capability to deal with trauma, surgical, critical care and medical evacuation needs.
- Approximately 75 Marines with six AH-1W Super Cobra and two UH-1N Huey helicopters from the Marine Medium Helicopter Squadron 264 (HMM-264) of the 26th MEU cross-decked from the USS *Iwo Jima*.
- Three HH-60H helicopters from Helicopter Anti-Submarine Squadron 3 (HS-3) cross-decked from the USS *Theodore Roosevelt* (CVN-71).

Initially, CTF-151 consisted of the *San Antonio*,

USS *Mahan* (DDG-72), and HMS *Portland* (F79), with additional warships expected to join this force.

On 28 January 2009, Japan announced its intention of sending a naval task force to join international efforts to stop piracy off the coast of Somalia. The deployment would be highly unusual, as Japan's non-aggressive constitution means Japanese military forces can only be used for defensive purposes. The issue has been controversial in Japan, although the ruling party maintains this should be seen as fighting crime on the high seas, rather than a "military" operation.

The process of the Prime Minister of Japan, Taro Aso, giving his approval is expected to take approximately one month. However, the Japanese Maritime Self-Defence Force (JMSDF) and the Japanese government face legal problems on how to handle attacks by pirates against ships that either have Japanese personnel, cargo or are under foreign control instead of being under Japanese control as current Article 9 regulations would hamper their actions when deployed to Somalia. It was reported on 4 February 2009, that the JMSDF was sending a fact-finding mission led by Gen Nakatani to the region prior to the deployment of the *Murasame*-class destroyer JDS *DD-106 Samidare* and the *Takanami*-class destroyer JDS *DD-113 Sazanami* to the coast of Somalia with a 13-man team composed of Japanese Ministry of Defence personnel, with members coming from the Ministry of Foreign Affairs and the JMSDF to visit Yemen, Djibouti, Oman, and Bahrain from February 8 to 20. Both JMSDF vessels are units of the 8th Escort Division of the 4th Escort Flotilla based in Kure, Hiroshima Prefecture.

The JMSDF's special forces unit, the Special Boarding Unit is also scheduled to potentially deploy to Somalia. The SBU has been deployed alongside the two destroyers to Somalia on 14 March 2009. According to JMSDF officials, the deployment would "regain the trust of the shipping industry, which was lost during the war." The JMSDF task force would be deployed in Somalia for 4 months. In their first mission, the *Takanami*-class destroyer JDS *DD-113 Sazanami* was able to ward off pirates attempting to hijack a Singaporean cargo ship. In addition, JMSDF P-3Cs are to be deployed in June from Djibouti to conduct surveillance on the

Somali coast. The House of Representatives of Japan has passed an anti-piracy bill, calling for the JMSDF to protect non-Japanese ships and nationals, though there are some concerns that the pro-opposition House of Councillors may reject it. The Diet of Japan has passed an anti-piracy law that called for JMSDF forces to protect all foreign ships traveling off the coast of Somalia aside from protecting Japanese-owned/manned ships despite a veto from the House of Councillors, which the House of Representatives have overturned. In 2009, the destroyers Harusame and DD-154 Amagiri left port from Yokusuka to replace the two destroyers that had been dispatched earlier on March 2009. Under current arrangements, Japan Coast Guard officers would be responsible for arresting pirates since SDF forces are not allowed to have powers of arrest.

The South Korean navy is also making plans to participate in anti-piracy operations after sending officers to visit the US Navy's 5th Fleet in Bahrain and in Djibouti. The South Korean cabinet had approved a government plan to send in South Korean navy ships and soldiers to the coast of Somalia to participate in anti-pirate operations. The ROKN was sending the *Chungmugong Yi Sun-sin class destroyer DDH 976 Munmu the Great* to the coast of Somalia. The Cheonghae Unit task force was also deployed in Somalia under CTF 151.

The Swiss government calls for the deployment of Army Reconnaissance Detachment operators to combat Somali piracy with no agreement in Parliament as the proposal was rejected after it was voted. Javier Solana had said that Swiss soldiers could serve under the EU's umbrella.

The Philippine government ordered the dispatch of a Naval Gunfire Liaison Officers to work with the US Navy's 5th Fleet as part of its contribution against piracy.

On 12 June 2009, Bulgaria also announced plans to join the anti-piracy operations in the Gulf of Aden and protect Bulgarian shipping, by sending a frigate with a crew of 130 sailors.

The Danish Institute for Military Studies has in a report proposed to establish a regionally-based maritime unit: a Greater Horn of Africa Sea Patrol, to carry out surveillance in the area to secure free navigation and take on tasks such as fishery inspection and environmental monitoring. A Greater Horn of Africa Sea Patrol would comprise elements from the coastal states-from Egypt in the north to Tanzania in the south. The unit would be established with the support of the states that already have a naval presence in the area.

In February 2010, Danish special forces from the *Absalon* freed 25 people from the Antigua and Barbuda-flagged vessel *Ariella* after it was hijacked by pirates off the Somali coast. The crew members had locked themselves into a store-room.

Southern African waters are becomingly an increasingly attractive alternative to the more protected Eastern African sea lanes. The recent rise in counter-piracy patrols is pushing more pirates down the coast line into unprotected areas of the Indian Ocean, which will require the joint navies' current patrols to widen their search area.

Somalia

Puntland

Between 2009 and 2010, the government of the autonomous Puntland region in northeastern Somalia enacted a number of reforms and pre-emptive measures as a part of its officially declared anti-piracy campaign. The latter include the arrest, trial and conviction of pirate gangs, as well as raids on suspected pirate hideouts and confiscation of weapons and equipment; ensuring the adequate coverage of the regional authority's anti-piracy efforts by both local and international media; sponsoring a social campaign led by Islamic scholars and community activists aimed at discrediting piracy and highlighting its negative effects; and partnering with the NATO alliance to combat pirates at sea. In May 2010, construction also began on a new naval base in the town of Bandar Siyada, located 25 km west of Bosaso, the commercial capital of Puntland. The facility is funded by Puntland's regional government in conjunction with Saracen International, a UK-based security company, and is intended to assist in more effectively combating piracy. The base will include a centre for training recruits, and a command post for the naval force. These numerous security measures appear to have borne fruit, as many pirates were apprehended in 2010, including a prominent leader. Puntland's security forces also reportedly managed to force out the pirate gangs from their traditional safe havens such as Eyl and Ṣar'ad, with the pirates now primarily operating from Hobyo, El Danaan and Harardhere in the neighboring Galmudug region.

Galmudug

Government officials from the Galmudug administration in the north-central Hobyo district have also reportedly attempted to use pirate gangs as a bulwark against Islamist insurgents from southern Somalia's conflict zones; other pirates are alleged to have reached agreements of their own with the Islamist groups, although a senior commander from the Hizbul Islam militia vowed to eradicate piracy by imposing sharia law when his group briefly took control of Harardhere in May 2010 and drove out the local pirates.

By the first half of 2010, these increased policing efforts by Somali government authorities on land along with international naval vessels at sea reportedly contributed to a drop in pirate attacks in the Gulf of Aden from 86 a year prior to 33, forcing pirates to shift attention to other areas such as the Somali Basin and the wider Indian Ocean.

Somaliland

The government of Somaliland, a self-declared republic that is internationally recognized as an autonomous region of Somalia, has adopted stringent anti-piracy measures, arresting and imprisoning pirates forced to make port in Berbera. According to officials in Hargeisa, Somaliland's capital, the Somaliland Coast Guard acts as an effective deterrent to piracy in waters under its jurisdiction.

Arab League Summit

Following the seizure by Somali pirates of an Egyptian ship and a Saudi oil supertanker worth $100 million of oil, the Arab League, after a meeting in Cairo, has called for an urgent summit for countries overlooking the Red Sea, including Egypt, Saudi Arabia, Sudan, Somalia, Jordan, Djibouti and Yemen. The summit would offer several solutions for the piracy problem, in addition to suggesting different routes and looking for a more secure passageway for ships.

Another possible means of intervention by the Red Sea Arab nations' navy might be to assist the current NATO anti-piracy effort as well as other navies.

United Nations

In June 2008, following the letter of the Transitional Federal Government to the President of the Council asking for assistance from the international community in its efforts to address acts of piracy and armed robbery against ships off the coast of Somalia, the UN Security Council unanimously passed a declaration authorizing nations that have the agreement of the Transitional Federal Government to enter Somali territorial waters to deal with pirates. The measure, which was sponsored by France, the United States and Panama, lasted six months. France initially wanted the resolution to include other regions with pirate problems, such as West Africa, but were opposed by Vietnam, Libya and most importantly by veto-holding China, who wanted the sovereignty infringement limited to Somalia.

The UN Security Council adopted a resolution on 20 November 2008, that was proposed by Britain to introduce tougher sanctions against Somalia over the country's failure to prevent a surge in sea piracy. The US circulated the draft resolution that called upon countries having naval capacities to deploy vessels and aircraft to actively fight against piracy in the region. The resolution also welcomed the initiatives of the European Union, NATO and other countries to counter piracy off the coast of Somalia. US Alternate Representative for Security Council Affairs Rosemary DiCarlo said that the draft resolution "calls on the secretary-general to look at a long-term solution to escorting the safe passage of World Food Programme ships." Even Somalia's Islamist militants stormed the Somali port of Harardheere in the hunt for pirates behind the seizure of a Saudi supertanker, the MV *Sirius Star*. A clan elder affiliated with the Islamists said "The Islamists arrived searching for the pirates and the whereabouts of the Saudi ship. I saw four cars full of Islamists driving in the town from corner to corner. The Islamists say they will attack the pirates for hijacking a Muslim ship." On 17 December 2008, the UN Security Council unanimously adopted a tougher resolution, allowing for the first time international land and sea occupations in the pursuit of pirates. Four ships, a Chinese fishing boat, a Turkish cargo ship, a Malaysian tug, and a private yacht were seized by pirates that same day. Resolution 1851 takes current anti-piracy measures a step further.

A Russian drafted resolution, Security Council Resolution 1918, adopted on 27 April 2010, called on all states to criminalise piracy and suggested the possibility of establishing a regional or international tribunal to prosecute suspected pirates.

Trials

Many of the suspects arrested in military operations in the Gulf of Aden in recent years have been set free for lack of evidence. Nearby countries in Africa have been reluctant to take on the burden of trials. In 2008, the Royal Navy was instructed by the Foreign Office not to arrest pirates for fear of breaching their human rights.

In May 2010, a Yemeni court sentenced six Somali pirates to death and jailed six others for 10 years each for hijacking a Yemeni oil tanker, killing one cabin crew member and leaving another missing in April 2009.

In May 2010, another Somali, Abdiwali Abdiqadir Muse, pleaded guilty in a New York federal court to seizing a United States-flagged ship *Maersk Alabama* and kidnapping its captain last year. He pled guilty, and was sentenced to 33 years imprisonment.

The first European trial of alleged Somali pirates opened in the Netherlands in May 2010. They were arrested in the Gulf of Aden in January 2009 when their high-speed boat was intercepted by a Danish frigate while allegedly preparing to board the cargo ship *Samanyolu*, which was registered in the Dutch Antilles. The pirates were sentenced to five years in prison, which was less than the maximum possible sentence. It is unlikely the men will be returned to Somalia after their sentence, as Somalia is considered too dangerous for deportation. One of the five has already applied for asylum in the Netherlands. Consequently, there are concerns that trials in European courts would encourage, rather than deter, pirates.

On 1 April 2010, the USS Nicholas (FFG-47) was on patrol off the Somali coast when it took fire from men in a small skiff. After chasing down the skiff and its mothership, US military captured five Somalis. Judge Raymond A. Jackson, a Federal District Court judge in Norfolk, Virginia threw out the piracy charge, which dates from enactment in 1819 when piracy was defined only as robbery at sea. The penalty for piracy is mandatory life in prison. The U.S. government is appealing the ruling. In March 2011 the five Somalis were sentenced to life for piracy to run consecutively with the 80-year term. In the same month 13 Somalis and one Yemeni suspected of hijacking and killing four Americans aboard a yacht made their first appearance in federal court in Norfolk.

On 28 January 2011, pursuant to the naval engagement of the pirate mother vessel MV Prantalay (a hijacked Thai trawler) by the INS Cankarso, the Indian Navy and the Indian Coast Guard killed 10 pirates and apprehended 15, while rescuing 20 Thai and Myanmarese fishermen being held aboard the ship as hostages. The rescued fishermen were sent to Kochi while the 15 pirates, of Somali, Ethiopian and Kenyan origin, have been taken to Mumbai. The Mumbai Police have confirmed that they have registered a case against the pirates for attempt to murder and various other provisions under the Indian Penal Code and the Passports Act for entering the Indian waters without permission.

The Somali government has questioned the authority of foreign countries to try the pirates abroad. The European Union has attempted to focus the prosecutions locally by involving Somalia's neighbors, but Somali authorities have called for the pirates to be tried at home.

Private Initiatives

There have been reports of pirates repelled by private initiatives. One such case would have occurred by the end of 2008, by armed personnel of transportation entrepreneur Barthe Cortes.

Other vessel owners and shipping line companies have also hired private security outfits for assistance. One such firm is Espada Logistics and Security Group based in San Antonio, Texas, whose security officers provide on-board protection from a ship's point of entry to its point of destination. They also offer anti-piracy training en route .to the Gulf of Aden, and have teamed up with African Shipping Lines, a leading international shipping line company, to provide security to vessels traveling along the coast of East Africa. Another private venture is MUSC, which specializes in counterpiracy and ship's security.

Proposed Solutions

In November 2008, the International Association of Independent Tanker Owners, a group of ship-owners representing 75% of the world's independent tanker fleet, asked for United Nations intervention. It called on the United Nations to co-ordinate anti-piracy patrols, and suggested the possibility of a naval blockade of Somalia and monitoring all vessels leaving the country's coastline. However, NATO responded by saying that it would be impossible to effectively blockade Somalia's vast coast. It also suggested that all home ports of Somali pirates be blockaded, or that ground forces be inserted in Somalia itself to destroy pirate bases.

Ultimately, many authors argue that the long-term solution to Somali piracy is political securitisation. Governments would have to employ socioeconomic measures such as poverty alleviation and good governance in order to deal with piracy (and even terrorism) effectively. In particular, a sustainable solution requires the establishment not only of effective governance but also the rule of law, reliable security agencies, and alternative employment opportunities for the Somali people. This however, would suggest possible military intervention, which there is a presence of animosity towards the idea since 1993's Operation Restore Hope.

Could Pirates Attack Your Cruise Ship? When Oceania Cruises announced this weekend that pirates had attempted to approach its Nautica as it cruised through the Gulf of Aden, it heightened existing concerns about passenger and cruise ship safety in that dangerous part of the world.

Nautica's captain, officers and crew successfully evaded the pirates yesterday, which is obviously the good news. No one was injured, and the ship is not even off schedule at this point. Still, increasing attacks by pirates on cargo vessels and oil tankers, not to mention this latest incident affecting a cruise ship, begs the question: If you're on a cruise ship that passes through the Gulf of Aden, what is the likelihood that pirates will attack? In the past three years, there have been three known incidents in which pirates have targeted cruise ships in the Gulf of Aden. In 2005, Seabourn Spirit was about 100 miles off the coast of Somalia when pirates fired AK-47 machine guns and rocket-propelled grenades in an attempt to highjack the ship. That failed. A more recent effort by pirates to attack Le Ponant — a French, luxury sailing vessel — in April 2008 was more successful. The ship (sailing crew-only) was hijacked; crewmembers were ultimately rescued, and the ship was recovered after a ransom was paid.

Compared to the sheer quantity of cruise and commercial ships moving through this area — and based on the fact that incidents impacting cargo vessels and oil tankers are clearly more common — these three incidents do not make as big of a splash. And the cruise lines, not to mention international maritime agencies, want to keep it that way.

You're never going to hear too much from cruise lines about their tactics involving security in the Gulf of Aden (or the Straits of Malacca, between Indonesia and Malaysia, which is another hot zone for pirates).

Aside from the obvious (why tip your hand to criminals), one reason is that there are no official regulations with which lines must comply. Case in point: last week, when Seabourn Spirit made its way through the Gulf of Aden, the cruise line decided to have the ship join a convoy, led by a French military vessel, as an extra safety precaution. It was the first time that Seabourn had joined such an effort, "and we changed our schedule to do so," said Bruce Good, Seabourn spokesman. The ship followed others along a protected course on the north side of the Gulf of Aden.

Oceania's Nautica traveled on its own, but according to cruise line spokesman Tim Rubacky, it did stick to the Maritime Safety Protection Area, which is patrolled by international anti-piracy task forces. What all lines tend to do consistently is gather intelligence — whether via government organizations, private security companies, military coalitions or the insurance companies that cover their ships. Some support is available for ship operators cruising in troubled waterways. Currently, NATO (North Atlantic Treaty Organization) has four ships patrolling the region (namely in the Gulf of Aden and on a route from Mombasa, Kenya, to Mogadishu and Somalia). The main purpose of these ships (operated by British, Turkish, Italian and Greek navies) is to escort vessels delivering supplies for the World Food Programme, but they also provide general anti-piracy patrols. The European Union will take over the job later this month.

Although there are no laws on the books to force shipping company compliance, the International Maritime Organization (IMO) does suggest tactics, techniques and technologies that offer interesting tips for pirate-attack evasion. For instance:

Ships should have a security plan that anticipates an attack by pirates. "Planning and training must be on the basis that an attack will take place and not in the belief that, with some luck, it will not happen." Officers and crew are asked to rehearse elements of the plan.

"Early detection of a possible attack is the most effective deterrent," the IMO guidelines stress. Tools to offer advance warning include, among others, "low-light binoculars," yacht radar that detects small crafts, night vision devices and barbed wire.

To avoid pirates, the IMO guidelines offer suggestions on evasion tools, ranging from relatively low-tech options to somewhat more sophisticated ones. On the common sense scale is the recommendation that "ships should travel with lights blacked out, except for mandatory navigation lights… in addition, turning on the ship's lights as attackers approach could alert them that they have been seen, dazzle them, and encourage them to desist."

One of the niftiest uses of technology to deter pirates, as employed by Seabourn Spirit in 2005, is the Long Range Acoustic Device (LRAD). Primarily used by warships, it serves as "a long-range hailing and warning, directed acoustic device, designed to communicate with authority and exceptionally high intelligibility in a 15-30 degree beam," Defence Update magazine said. It's, in essence, a device that sends sonic waves out to a target, and the resulting din can cause eardrums to pop and bleed, which, in effect, stuns the criminals.

One Cruise Critic member who traveled on Holland America through the Gulf of Aden reported seeing water hoses laid out on open decks, and the IMO report notes that water pressure of 80 pounds per square inch can deter attackers. Another tactic? "Provided that navigational safety allows, masters should consider 'riding off' attackers' craft by heavy wheel movements...the effect of the bow wave and swash may deter would-be attackers."

Interestingly, ships' officers are not encouraged to carry firearms. The rationale? "Carriage of arms on board may encourage attackers to carry firearms, thereby escalating an already dangerous situation."

One of the major differences between commercial vessels and cruise ships is that, for the former, leaving the region — with its proximity to the Suez Canal, which connects the east to the west — is not a possibility. For cruise lines? The reality, said Seabourn's Good, is that "the impact of continued predation by these pirates has a much bigger effect on the shipping economy than on us."

Still, if enough passengers vote with their wallets to skip the repositioning cruises that transit ships from the Mediterranean to Asia, it could impact deployment. "One of the basics of deployment is that you put ships where people will buy tickets," Good said. If the stakes get riskier — and if passengers are increasingly put off — you could see a real drop in opportunities to cruise in such diverse and exotic places as the Middle East and Asia.

Piracy, Law of the Sea, and Use of Force: Developments off the Coast of Somalia

The Revival of Piracy and Other Violent Acts off the Coast of Somalia

Although never absent from the international scene – one may recall attacks on ships carrying 'boat people' off the coasts of South-East Asia – pirates seemed to have ceased to be a general menace to the international community justifying the traditional qualification of *hostis humani generis,* until the massive development of their activities off the coasts of Somalia since 2000, and, especially, 2006.

Capturing ships and holding them and their crews for ransom since the 1990s has been carried out by armed groups acting mostly in the territorial sea and claiming to protect Somalia's fishing resources, which were in effect pillaged by foreign fishermen, and the coastal waters, which were used as a dumping ground for waste in the absence of a government able to enforce the law. Taking advantage of the continuing lack of an effective government, and not without connection with terrorist groups and with the political and armed fights going on in Somalia, pirate activity then absorbed a growing number of people – including fishermen expert in handling boats – and became ever bolder. It now represents a very serious menace to navigation coming from the Suez Canal and going through the Gulf of Aden to the narrow area between the Horn of Africa and the Arabian peninsula. In these sea areas off the Somali coast, as well as in those south of the Horn of Africa, piracy has developed, attacking ships even at a great distance from the coast. The success in capturing ships and crews and in obtaining substantial amounts of money as ransom, as well as their efficient way of dealing with money so obtained, have again made pirates, *sub specie* of the Somali pirates, the *hostes humani generis.* The danger to navigation through a choke-point of international traffic, as well as the outrage aroused by pirate attacks on ships carrying humanitarian supplies to the Somali population, have been decisive in alarming states all over the world.

The Security Council has linked the activities of pirates off the coast of Somalia with the notion of a threat to international peace and security. Since Resolution 733/1992, the Security Council has routinely invoked Chapter VII as regards the situation in Somalia, and stated that such situation constitutes or continues to constitute 'a threat to international peace and security', in its first resolution on piracy off the coasts of Somalia, 'determine[d]' that such piracy 'exacerbate[s] the situation in Somalia which continues to constitute a threat to international peace and security in the region'. The declaration made in approving Resolution 1851 by the Chinese Minister of Foreign Affairs at the Security Council meeting held at the level of Foreign Ministers on 16 December 2008 clearly shows this approach: '[t]he long-term delay in the settlement of the Somali issue is posing a serious threat to international peace and security, while the rampant piracy off the Somali coast has worsened the security situation in Somalia'. The link is made indirectly, avoiding the criticism which the Council often incurs when applying this notion to matters hitherto not considered to be covered by the notion of threat to international peace and security. It nonetheless achieves the objective that action against piracy off the Somali coasts be conducted within the framework of Chapter VII of the UN Charter.

The Law of the Sea Rules on Piracy and their Inadequacy to Cope with the Violent Activities off the Somali Coast

The international law of piracy is set out in Articles 100 to 107 and 110 of the UN Convention on the Law of the Sea (UNCLOS). The fact that these Articles repeat almost literally Articles 14 to 22 of the Geneva Convention on the High Seas of 1958, and that some states, including the United States as well as Israel, Switzerland and Venezuela, while not bound by UNCLOS, are bound by the Geneva Convention, entails that, as a matter either of customary or of conventional law, these Articles state the law as currently in force.

For present purposes it seems necessary and sufficient to recall the provisions concerning the definition of piracy and action against pirates.

As regards the definition, its essential aspect is that piracy consists of 'any illegal acts of violence or detention, committed for private ends by the crew or the passengers of a private ship or aircraft and directed ... on the high seas against another ship or aircraft, or against persons or property on board such ship or aircraft'.

As regards action which may be taken against a pirate ship, apart from the right of warships of all states to exercise the right to visit aimed at ascertaining whether a ship is engaged in piracy, the main provision is Article 105, which states: On the high seas, or in any place outside the jurisdiction of any State, every State may seize a pirate ship or aircraft, or a ship or aircraft taken by piracy and under the control of pirates, and arrest the persons and seize the property on board. The courts of the State which carried out the seizure may decide upon the penalties to be imposed, and may also determine the action to be taken with regard to the ships, aircraft or property, subject to the rights of third parties acting in good faith.

The definition of piracy is rather narrow, as it includes only action on the high seas and only action undertaken by one ship against another ship. So forms of violence conducted in the territorial sea as well as without the involvement of two ships, such as, for instance, the violent taking of control of a ship by members of its crew or passengers, even when the follow-up consists of holding to ransom the ship and its crew and passengers, are not included.

Correctly, the taking of control by hijackers embarked as passengers on the Portuguese ship *Santa Maria* in 1961 and on the Italian cruise ship *Achille Lauro* in 1985, which had extensive press coverage, were not considered to be piracy. Violent activities against ships off the Somali coast sometimes take place in whole or in part in the territorial seas, thus often remaining outside the scope of the definition. More rarely they do not involve the presence of one or more other ships, as usually very fast skiffs are used, coming from bases on the mainland or from 'mother ships' at sea. It may be underlined that acts preparatory to piracy and other acts of violence not directly linked to piracy are not included in the definition.

As far as the action to be taken is concerned, under Article 105 the flag state of the seizing ship enjoys very broad powers. These consist of the right to arrest persons and to seize property, and, through the above mentioned rights, to decide upon penalties and on action to be taken with regard to the ship, aircraft and property, the right to submit the persons arrested and the property seized to judicial proceedings. In other words, the universal jurisdiction of the seizing state's courts is supported by international law. The language of Article 105 ('may') seems to indicate that the exercise of jurisdiction by the seizing state's courts is a possibility, not an obligation, notwithstanding the 'duty' to cooperate in the repression of piracy set out in Article 100. The rule in Article 105 does not, however, establish the exclusive jurisdiction of the seizing state's courts. Courts of other states are not precluded from exercising jurisdiction under conditions which they establish. Thus the international law rules on action to be taken against pirates permit action, but are far from ensuring that such action is effectively taken.

The Security Council Resolutions' Broadening of the Scope of International Law Rules on Piracy

With Resolution 1816 of 2 June 2008 and the others which followed it, especially Resolutions 1846 of 2 December 2008 and 1851 of 18 December 2008, the Security Council has endeavoured to cope with the growing alarm caused by pirate activities off the coast of Somalia. It has taken measures within the framework of Chapter VII which aim at remedying the limitations of the above mentioned rules of international law, as far as their application to the situation at hand is concerned.

These Resolutions, while using the term 'piracy', do not define it. References to the provisions of UNCLOS and the statement that these provisions 'provide guiding principles for cooperation to the fullest possible extent in the repression of piracy' indicate that the starting point is the above-recalled definition in the Convention. These Resolutions, however, always mention 'armed robbery' together with piracy. Armed robbery is not defined. It is a term routinely used within the framework of IMO, and may be understood to include all acts of violence the purposes of which are identical or similar to those of piracy but are not covered by the conventional definition of it, in particular because they may be perpetrated without using a ship against the target ship. In IMO parlance 'armed robbery', however, refers only to activities in waters under the jurisdiction of a state, so that it does not extend the scope of provisions on piracy to acts committed on the high seas unless two ships are present. This is what the Security Council Resolutions do, as they use the expression 'piracy and armed robbery against vessels in the territorial waters of Somalia and the high seas off the coast of Somalia'. As two or more ships are involved in most of the Somali cases, the mention of 'armed robbery' would seem not to be strictly dictated by the needs of existing practice, and rather inspired by the aim of including all acts connected with piracy (such as preparatory acts) and future possible acts involving only one ship.

The key element in the Resolutions is set out in paragraph 7 of Resolution 1816. It copes with the limitation of the definition of piracy to acts perpetrated on the high seas which, as mentioned, makes it inadequate to deal with acts which sometimes take place wholly in the territorial sea, and very often include an attack on the high seas followed by the pirated ship being brought by the pirates into the territorial sea and held for ransom in a port or near the coast, or by the attacking skiffs' retreat into the territorial and internal waters of Somalia.

This key element is that certain states (to which I will come later) are authorized to:

- (a) Enter the territorial waters of Somalia for the purpose of repressing acts of piracy and armed robbery at sea, in a manner consistent with such action permitted on the high seas with respect to piracy under relevant international law; and
- (b) Use, within the territorial waters of Somalia, in a manner consistent with action permitted on the high seas with respect to piracy under relevant international law, all necessary means to repress acts of piracy and armed robbery.

The basic effect of these provisions is to make the rules of international law concerning piracy on the high seas applicable also to territorial waters, *inter alia* permitting pursuit from the high seas into these waters, and clarifying that states acting under these rules within the territorial waters of Somalia may use 'all necessary means'.

It may be added that – following an episode in which French troops pursued pirates into the Somali mainland – Resolution 1851 added on 16 December 2008 an authorization to conduct 'all necessary measures that are appropriate *in Somalia* for the purpose of suppressing acts of piracy and armed robbery at sea'. The expression 'in Somalia', while not explained in the preambular paragraphs, clearly alludes to action undertaken on the mainland.

The Limitations of the New Rules: **Ratione Temporis, Ratione Loci***; the Concern about Changing International Customary Law*

Although the main effect of Resolution 1816 and the following ones is to extend both *ratione loci* and *ratione materiae* the scope of the international law rules concerning piracy, the Security Council has framed the relevant resolutions very cautiously. It has introduced a number of limitations which make the provisions adopted less revolutionary than they might appear, and seem aimed, in particular, at fending off possible criticism of the Council acting as a 'legislator'.

First, the authorization given is limited *ratione temporis.* Resolution 1816 limits to six months the validity of the authorization it introduces, while providing for a progress report and a more complete report on the application of the resolution to be submitted within, respectively, three and five months and stating the intention to review the situation and consider, 'if appropriate, renewing the authority provided in paragraph 7 for additional periods'. The authority has in fact been renewed for a period of 12 months by Resolution 1846 of 2 December 2008. The authorization to undertake all necessary measures ''in Somalia' set out in Resolution 1851 is also limited to the 12 months starting with the adoption of Resolution 1846.

Second, the scope of the resolutions is clearly limited *ratione loci* as it is stated that the authorization provided 'applies only with respect to the situation in Somalia'. This implies in particular that authorization to enter the territorial sea does not apply to the territorial sea of states other than Somalia (such as Yemen or Kenya).

Third, the resolutions request that activities undertaken pursuant to the authorizations they set out 'do not have the practical effect of denying or impairing the right of innocent passage to the ships of any third State'. This provision seems consistent with the idea that, while the authorizations set out in the resolutions introduce a limitation to the sovereignty of the Somali coastal state on its territorial sea, they should not have any effect on the rights third states (states other than the coastal and authorized states) are entitled to exercise in the territorial sea, such as, especially, the right of innocent passage.

Fourth, the resolutions affirm that the authorization they contain 'shall not affect the rights obligations or responsibilities of member States under international law, including any rights or obligations under the [Law of the Sea] Convention with respect to any other situation' and underscore in particular that they 'shall not be considered as establishing customary international law'.

These provisions correspond to concerns firmly stated in the Security Council by representatives of developing states keen to maintain the integrity of the UN Law of the Sea Convention. Of course, it cannot be ruled out that, if authorizations similar to those granted as regards the Somali situation were to be routinely granted in other situations, the possible formation of a customary rule could be at least discussed. The provisions just quoted would provide, however, a strong, although perhaps not insurmountable, argument against.

A comparison between the Security Council resolutions and the non-binding Code of Conduct adopted on 29 January 2009 concerning the Repression of Piracy and Armed Robbery against Ships in the Western Indian Ocean and the Gulf of Aden shows how keen many states are not to go beyond what is provided in UNCLOS and how dangerous the provisions of the Security Council resolution may appear to them. The Code provides different rules for piracy on the high seas and for armed robbery in internal, archipelagic, and territorial waters. On the high seas the UNCLOS regime applies, and in the territorial sea, including pursuit from the high seas, the authorization of the coastal state is necessary.

The Requirement of the Consent of the TFG

The Security Council resolutions considered here are adopted on the basis of the authorization of the Somali Transitional Federal Government (TFG). Paragraph 9 of Resolution 1816 'affirms' that the authorization set out in paragraph 7 'has been provided only following receipt of the letter from the Permanent Representative of the Somalia Republic to the United Nations to the President of the Security Council dated 27 February 2008 conveying the consent of the TFG'.

Similar formulations, referring to further letters conveying the consent of the TFG, are in Resolution 1846 and 1851. From the tenor of a number of declarations made upon the adoption of the resolutions, it would seem that without such authorization, and notwithstanding the lack of control by the TFG of the waters off Somalia, unanimity in the Security Council would not have been reached. The reference to the authorization of the coastal state takes away all, or much of, the revolutionary content of the resolutions. Indeed, the activities purportedly 'authorized' by the Security Council in light of the coastal state's authorization could also be conducted in the absence of a Security Council resolution adopted within the framework of Chapter VII.

Under international law, states are free to dispose of their rights in their territorial seas, for instance by allowing other states to conduct police activities in them. A precedent which may be quoted is the exchange of Notes of 25 March 1997 between Albania and Italy, in which Albania agreed that Italian naval forces could in Albanian territorial waters stop ships flying whatever flag and carrying Albanian citizens which had evaded controls exercised by the authorities of Albania in the latter's territory.

The fact that no authorization by the Security Council under Chapter VII to exercise jurisdiction in the territorial sea of a state is needed if there is an authorization of the coastal state is confirmed by the language used by the EU Council's Joint Action concerning 'Operation Atalanta' in the waters off Somalia.

The provision concerning transfer, for the purpose of the prosecution of arrested pirates or armed robbers, is set out 'on the basis of Somalia's acceptance of the exercise of jurisdiction by Member States or by third State, on the one hand [namely, as regards the territorial sea], or of article 105 of the United Nations Convention on the Law of the Sea, on the other [namely, as regards the high seas]'. To invoke Security Council resolutions as a basis has apparently (and correctly) been seen as superfluous.

The importance of the coastal state's consent seems highlighted by the fact that, contrary to the international law rules on piracy on the high seas which permit seizure of pirate ships by 'every State'. the Security Council resolutions limit the authorizations they provide to 'States cooperating with the TFG' for which 'advance notification has been provided by the TFG to the Secretary General'. Thus, the coastal state maintains (in fact, is allowed to maintain) control as regards which states are authorized to enter its territorial sea and, indeed, territory, to fight pirates and armed robbers. At present it seems that the flotilla patrolling the waters off the coast of Somalia (not necessarily its territorial sea) includes naval forces of about 20 states, coordinated by the United States.

For the first time China has deployed naval vessels outside the seas adjacent to it, and the European Union has formed and deployed a joint naval force within 'Operation Atalanta'.

The importance given to the coastal state's consent, unnecessary for action under Chapter VII, seems to pursue three objectives. The first is to pay homage to state sovereignty, meeting the above mentioned concerns that through these resolutions new customary international law rules could be 'established'. The second is to strengthen the TFG, which, while maintaining the Somali presence at the United Nations, does not exercise effective power in Somalia, and in particular lacks the capacity to fight pirate activities off its coasts.

The third, through the designation by the TFG of the states whose vessels are authorized to act in its territorial sea, would seem to consist in limiting the foreign fleets' presence in Somali waters to those of the states most involved, and to states ready to cooperate with each other.

The overlap between the Security Council's authorization and that of the coastal state as represented by the TFG may, however, serve, additionally, or perhaps especially, another purpose. It must be recalled that, by a law of 1972, Somalia adopted a territorial sea of 200 miles width and that, although Somalia ratified UNCLOS on 24 July 1989, there is no record (at least available to the present writer) that that law has been revoked.

In the situation of possible conflict between a domestic statute and the international obligation assumed under UNCLOS to have 12 miles as the maximum width for the territorial sea, compounded by the lack of effective authority in Somalia, the intent of the Security Council in ensuring the TFG's consent to action by other states against pirates and armed robbers within the territorial sea may be explained as to grant a legal basis for such action whatever the width of the Somali territorial sea. From the behaviour of states patrolling the waters off the coast of Somalia it would seem clear that they assume that the external limit of the Somali territorial sea is 12 miles. Whether this is also the assumption of the TFG is uncertain, and the permission to act against pirates and armed robbers in its territorial sea has the beneficial result of avoiding discussion on this question.

What to Do with Captured Pirates and Armed Robbers? Human Rights, Agreements with Neighbouring States, the SUA Convention

As mentioned, international law accords universal jurisdiction to the courts of the seizing state. This jurisdiction, applicable under Article 105 of UNCLOS for the seizure and arrest of pirates on the high seas, applies also to seizures and arrests in the territorial sea of Somalia under the Security Council resolutions referred to above.

The seizing states – in other words, the states fighting pirates and armed robbers in the waters off Somalia and having arrested them – are, however, reluctant to exercise such broad powers by prosecuting and submitting to criminal proceedings in their courts the pirates and armed robbers arrested. They seem concerned by the expense involved, by legal complexities, relating for instance to evidence, inherent in criminal proceedings to be held far away from the place where the alleged crime was committed, and, perhaps especially, by the human rights implications of exercising jurisdiction.

A recent case highlights these difficulties. The Danish Navy ship *Absalon* on 17 September 2008 captured 10 pirates in the waters off Somalia. After six days' detention and the confiscation of their weapons, ladders, and other implements used to board ships, the Danish government decided to free the pirates by putting them ashore on a Somali beach. The Danish authorities had come to the conclusion that the pirates risked torture and the death penalty if surrendered to (whatever) Somali authorities. This was unacceptable, as Danish law prohibits the extradition of criminals when they may face the death penalty. Moreover, they were not ready to try them in Denmark as it would be difficult (in light of the possible abuses they would risk) to deport them back to Somalia after their sentences were served. It is clear that human rights considerations, or perhaps reasons of expediency presented as human rights concerns, prevailed over considerations concerning the fight against piracy. In the same vein, the British Foreign Office reportedly warned the Royal Navy against detaining pirates since this might violate their human rights and could lead to claims to asylum in Britain.

The capture and detention at sea of criminals later brought to trial in far away courts of the state of the arresting vessel have been referred by those captured to the judgment of the European Court of Human Rights (ECtHR). This happened in the *Rigopoulos* and *Medveyev* cases decided on 12 January 1999 and 10 July 2008 respectively. In these cases the applicants were arrested on the high seas on a ship boarded, with the authorization of the flag state, under suspicion, later proved to be well founded, of being engaged in smuggling narcotic drugs. The question submitted to the Court was whether detention on the arresting naval vessel for about two weeks was compatible with Article 5(3) of the European Convention on Human Rights according to which, *inter alia*, arrested or detained persons 'shall be brought promptly before a judge or other officer authorized by law to exercise judicial power'. The Court, even though in both cases it decided that the circumstances were exceptional enough to justify an affirmative answer, stated clearly that the principle was that such a long period of detention was not compatible with the provision in question. Consequently, states parties to the European Convention may, in different circumstances, be confronted by a decision finding a violation of the human rights of the detained criminal (be it a drug trafficker or a pirate). Moreover, in the 2008 *Medveyev* case the ECtHR found a violation of Article 5(1) of the European Convention, according to which *inter alia*, '[n]o one shall be deprived of his liberty save in the following cases and in accordance with a procedure prescribed by law' on the basis of a strict interpretation of the agreement authorizing the boarding and arrest, but allegedly not sufficiently clear as to the right to submit the arrested person to trial. The possibility of a similar decision seems highly unlikely in the case of piracy, in light of the broad powers recognized by general international law and the Security Council resolutions. It shows, nevertheless, that a court like the ECtHR will tend to interpret the law of the sea and international law rules in such a way as to offer maximum protection to the individuals involved.

The reluctance of seizing states to prosecute and try pirates is implicitly taken into account in the resolutions of the Security Council. Resolutions 1816 and 1846 contain the following provision :*Calls upon* all States, and in particular flag, port and coastal States, States of the nationality of victims and perpetrators or piracy and armed robbery, and other States with relevant jurisdiction under international law and national legislation, to cooperate in determining jurisdiction, and in the investigation and prosecution of persons responsible for acts of piracy and armed robbery off the coast of Somalia, consistent with applicable international law including international human rights law, and to render assistance by, among other actions, providing disposition and logistics assistance with respect to persons under their jurisdiction and control, such victims and witnesses and persons detained as a result of operations conducted under this resolution.

This language is merely hortatory. However, most states involved as flag states of the ships which are victims of piracy or of the ships patrolling the waters off Somalia, or as neighbouring coastal states such as Djibouti, Kenya, and Yemen, but not Somalia, are bound by the precise obligations set out for parties to the Rome Convention for the Suppression of Unlawful Acts Against the Safety of Maritime Navigation of 10 March 1988 (the SUA Convention). This Convention – adopted in the wake of the *Achille Lauro* affair – provides that states parties shall establish a number of criminal offences, most of which correspond in whole or in part with actions committed by pirates or armed robbers (notably, it does not require the presence of two ships and does not distinguish between maritime areas). In particular, it makes it compulsory to 'take such measures as may be necessary to establish jurisdiction' over such offences for the flag state of the ship against or on board which the crime is committed, for the state in the territory of which, including the territorial sea, the crime is committed, and for the state the national of which committed the offence. It further authorizes other states to establish jurisdiction in additional cases; these include the state of which a person seized, threatened, injured, or killed is a national and the state which the criminal act is intended to compel to do or abstain from doing any act. In all cases it is compulsory for the state in the territory of which the alleged offender is present to establish jurisdiction, and it does not extradite such offender to one of the states which has established jurisdiction.

While the SUA Convention is not mentioned in Resolution 1816, Resolution 1846 recalls the obligations set out in it and urges states parties 'to fully implement' these obligations.

The reluctance of the seizing states is not the only cause of lack of efficiency in dealing with captured pirates and armed robbers. As was said in a preambular paragraph of Resolution 1846, 'the lack of capacity, domestic legislation, and clarity about how to dispose of pirates after their capture has hindered more robust international action against the pirates off the coast of Somalia and in some case led to the pirates being released without facing justice'.

In particular, the difficulties met in surrendering captured pirates to a neighbouring state and ensuring that such state will exercise jurisdiction and respect human rights have led to bilateral agreements, mentioned in the media but not made public, between the United Kingdom and the United States and Kenya. According to a declaration of the Kenyan Foreign Minister, these are 'memoranda of understanding' which will not, however, 'be an open door for dumping pirates onto Kenyan soil'. They will function on a case-by-case basis. These memoranda seem to be part of a practice, mentioned and encouraged by Resolution 1846, consisting in agreements between the states whose ships are patrolling the waters off the coast of Somalia and 'countries willing to take custody of pirates' to place on such ships law enforcement officers (called 'shipriders') 'from the latter countries, in particular countries of the region, to facilitate the investigation and prosecution of persons detained'. These arrangements require the prior consent of the TFG 'for the exercise of third State jurisdiction by shipriders in Somali territorial waters' and provided they 'do not prejudice the effective implementation of the SUA Convention'.

While the practice of shipriders under these agreements permits the jurisdiction of Kenya (or of other states whose officers are on board the seizing vessels) to be based on the fiction that it is a seizing state, in other cases the connecting factor is the nationality of the victim ship or of the victims. So reportedly the Netherlands has signed an agreement with Denmark to extradite five Somali pirates who attacked a Netherlands-Antilles cargo vessel in the Gulf of Aden and were captured by Denmark.

European Union Joint Action 2008/851/CSFP on the EU military operation against piracy and armed robbery off the Somali coast envisages the situation now considered. It provides that, if the competent authorities of the flag state of the ship 'which took them captive' 'cannot or do [] not wish to exercise its jurisdiction', the persons arrested during the operation with a view to their prosecution shall be transferred to a 'Member State or any third State which wishes to exercise its jurisdiction over the aforementioned persons'. This provision does not rule out the use of 'shipriders', but adds possible transfer to other EU Member States or to other willing third states. In light of the above mentioned concerns, it seems relevant to quote a provision of the same Joint Action stating that:no person...may be transferred to a third State unless the conditions for the transfer have been agreed with that third State in a manner consistent with relevant international law, notably international law of human rights, in order to guarantee in particular that no one shall be subjected to the death penalty, to torture or to any cruel, inhuman or degrading treatment.

Detailed provisions for the protection of the human rights of pirates and armed robbers captured by the European Union naval force and transferred to Kenya are set out in articles 3 and 4 of the exchange of letters made in Nairobi on 6 March 2009 on such transfer concluded between the European Union and Kenya. It is to be noted that this agreement, though not ruling out their involvement, does not mention 'shipriders'.

Use of Force in Operations at Sea against Pirates and Armed Robbers

Seizing a pirate ship under the power granted to all states by UNCLOS implies the possibility of the use of force. This is even clearer under the above resolutions of the Security Council which mention the use of 'all necessary means for repressing acts of piracy and armed robbery'. It is well known that in the parlance of the Security Council 'all necessary means' means 'use of force'. The above-quoted EU Council Joint Action makes this explicit in defining the mandate of 'Operation Atalanta' when it says that Atalanta shall take 'all necessary measures, including use of force'.

This is not use of force against the enemy according to the law of armed conflict, because there is no armed conflict, international or internal. Pirates are not at war with the states whose flotillas protect merchant vessels in the waters off the coast of Somalia. It has been argued that pirates not being combatants are civilians who, under international humanitarian law, may not be specifically targeted except in immediate self-defence. Whatever opinion one holds about the applicability of the law of armed conflict, it is a fact that practice in the waters off Somalia seems to indicate that warships patrolling these waters resort to the use of weapons only in response to the use of weapons against them. So in an incident in the Gulf of Aden reported on 14 November 2008, a British naval vessel having positively recognized a Yemeni cargo ship which had participated in a hijacking attempt on a Danish cargo ship on the same day, tried to stop it by 'non forcible methods'. Only when these had failed, did 'the Royal Navy launch [] small assault craft to encircle the vessel'. Once the pirates opened fire, 'the Navy fired back in self-defence'. In another episode reported on 21 November 2008, the Indian Navy vessel *Tabar* patrolling the Gulf of Aden 285 miles off the coast of Oman requested a vessel described as a pirate mother ship, the crew of which was seen 'with a full complement of modern weapons', to stop. When the pirate ship 'responded by threatening to "blow up the naval warship if it closed on her"' and fired at the Indian vessel, the *Tabar* responded and sank the vessel.

Thus, self-defence against an armed attack or the threat thereof, either in the questionable framework of the law of armed conflict or in the discussed framework of resort to it against non-state actors, or, more likely, as a self-imposed rule of engagement for police action, seems to be a guiding principle of states the navies of which are engaged in fighting pirates off the coast of Somalia and neighbouring states.

The question must, however, be raised whether force may be used in action against pirates and armed robbers independently of self-defence, and whether, if an affirmative answer is given, international law prescribes limits to such use. Action against pirates may, in my view, be assimilated to the exercise of the power to engage in police action on the high seas on foreign vessels which is permitted to other states by exceptions to the rule affirming the exclusive jurisdiction of the flag state. Such permission is rarely and reluctantly given by flag states unless upon request on a case-by-case basis. There are nonetheless examples of permission given in general terms in a few treaties on drug trafficking and fisheries. Among these, the 1995 UN Fish Stocks Agreement is the main multilateral instrument. This agreement permits certain non-flag states to board and inspect fishing vessels on the high seas. In principle such action should not involve the use of force, as the flag state is, *inter alia,* bound to 'accept and facilitate prompt and safe boarding by the inspectors' and sanction the master if he refuses to agree to being boarded. The possibility of the use of force after the boarding is nevertheless envisaged in Article 22(1)(f), which states that the inspecting state shall 'avoid the use of force except when and to the degree necessary to ensure the safety of the inspectors and where the inspectors are obstructed in the execution of their duties'. General international law, in authorizing stopping and boarding for the purpose of exercising the right of visit under Article 110 of UNCLOS or the seizure of a pirate ship under Article 105, presupposes that force may be used to reach these objectives. In view of the fact that they have accepted the relevant instruments and are bound by the relevant customary rules and by the relevant resolutions taken by the Security Council under Chapter VII, states can be considered as consenting to, or as being obliged to accept, the use of force undertaken in order to undertake these police activities.

Limits to such use of force in the exercise of police action authorized by international law have been indicated in dispute-settlement and treaty practice. Repeating and developing points made in the *I'm Alone* arbitration award and in the report of the Commission of enquiry on the *Red Crusader* case, the International Tribunal for the Law of the Sea, in its *M/V Saiga No 2* judgment states: 'international law ... requires that the use of force must be avoided as far as possible and, where force is inevitable, it must not go beyond what is reasonable and necessary in the circumstances. Considerations of humanity must apply in the law of the sea, as they do in other areas of international law.' The judgment further recalls the practice concerning visual and auditory signals to stop, firing shots across the bows, and a variety of other measures, normally followed before resorting to force. In the same vein, the UN Fish Stocks Agreement concludes Article 22(1)(f) by stating, '[t]he degree of force used shall not exceed that reasonably required in the circumstances'.

The *Saiga No 2* case was quoted in the 2007 Arbitral award deciding a maritime border dispute between Guyana and Suriname. The question addressed was whether the order given by a Surinamese naval vessel to a Guyanese oil drilling platform situated on a disputed area of the continental shelf to leave the area together with the statement that 'if they would not do so, the consequences would be theirs' constituted an illegal threat of use of force under Article 2(4) of the UN Charter and under customary law. The Arbitral Tribunal considers *inter alia* the remarks set out in the ICJ judgment in 1986 in *Military and Paramilitary Activities in and against Nicaragua* distinguishing the gravest forms of use of force which constitute an armed attack from 'other less grave forms'. While not directly drawing consequences from this distinction, it 'accepts the argument that in international law force may be used in law enforcement activities provided that such force is unavoidable, reasonable and necessary' but concludes that, 'in the circumstances of the present case, ... the action mounted by Suriname on 3 June 2003 seemed more akin to a threat of military action rather than a mere law enforcement activity', and that it contravened the UN Charter and general international law. The Tribunal does not specify whether this conclusion is based on the characteristics of the threatened use of force (possible involvement of military means?) or on the fact that the use of force was illicit because it had taken place in an area the Award had determined to belong to Guyana and where consequently Suriname had no enforcement rights, and that it would have been considered a legal enforcement activity had the area been found to belong to Suriname. The latter seems to be the most reasonable explanation, but, admittedly, the Award is not clear on this key point.

As regards the use of force against pirates, the *Guyana v. Suriname* award seems to confirm the emerging trend that activities permitted by international law for the enforcement of rights may include the use of force, provided such force is unavoidable, reasonable, and necessary. The practice seen above clarifies these requirements by introducing the element of respect for the human rights of the persons involved which was implicit in the mention of 'considerations of humanity' in the *M/V Saiga No. 2* judgment of the International Tribunal for the Law of the Sea.

Ending Somalian Piracy: Pitfalls and Possibilities of Australian Naval Intervention and Long-term Human Security Policy Initiatives

Carolin Liss of Murdoch University discusses "the possible benefits and disadvantages of Australian participation in an international, UN approved, anti-piracy task–force operating in the Gulf of Aden and off the Horn of Africa. In themselves, Liss notes, "the patrols of pirate-infested waters are of limited use because they will only foil individual attacks, and only within the area covered by patrol vessels." As a result, Liss argues, "countries such as Australia, Japan and South Korea that consider sending warships to Somalia therefore have to understand that deploying warships could be a long-term commitment that may make little difference – particularly in regard to human security – if the political and humanitarian situation in Somalia is not addressed.". In conclusion Liss notes that "the deployment of an Australian warship would make little sense unless Australia and other countries also take an active role in ensuring that long-term measures are taken and successfully implemented."

The world's most blatant pirate attacks are currently taking place off the coast of Somalia. In the past 12 months, more than 100 ships were attacked in this dangerous area. These attacks included more than 40 hijackings of merchant and fishing vessels, with the pirates receiving millions of US dollars in ransom money for kidnapped crew and hijacked vessels. When attacks on vessels off the coast of Somalia became more frequent and serious in nature, international concern about the safety of ships and crews passing the Horn of Africa grew. As a result, nations from around the world have sent warships to combat piracy in the area. In January 2009, Australian Defence Force Chief, Air Chief Marshal Angus Houston, voiced the possibility of Australian involvement in the fight against Somali pirates. This paper discusses the possible benefits and disadvantages of Australian participation in an international, UN approved, anti-piracy task–force operating in the Gulf of Aden and off the Horn of Africa

The Problem: Piracy in Somalia

The number and scale of pirate attacks currently occurring in the Gulf of Aden and off the Horn of Africa are unprecedented and pose a serious threat to international shipping. Over the past several months, pirates have attacked UN aid ships, hijacked merchant and fishing vessels, held crew hostage and collected millions in US dollars in ransom payments. Most attacks occurred in the Gulf of Aden and off the coast of Somalia, with pirate gangs using mother ships to conduct attacks too far from the coast for speedboats to reach. At the start of 2009, pirates in the area hold up to 15 ships at a time while ransom negotiations between shipowners and pirates take place. The negotiation processes can take months, with the crew held hostage but generally treated well. Once a ransom is paid, the crew and vessel are released and the pirates hijack a new vessel.

With their newfound wealth, the pirates are believed to buy additional boats and weapons for future attacks and to have built houses and bought expensive cars. Shops catering to the pirates have sprung up along the coast in areas where hijacked ships are held, providing supplies for the pirates and hostages. Over the past several weeks some pirates have freely given interviews to the international media, explaining their motives and aims. Their activities, they claim, started as a protective measure against illegal fishing in their waters, which depleted fish stocks in the area, robbing local fishers of their income and livelihood. Over time their activities changed and more and more international merchant vessels were targeted after substantial ransoms were paid. According to the pirates, their activities are driven by poverty, with some perpetrators expressing the hope that their attacks on international shipping will attract attention to the poverty and conflict in Somalia and the suffering of the local population. They stress, however, that they are interested in ransom payments only and have so far not voiced political demands in exchange for hijacked ships, except for the release of captured pirates.

Among the vessels attacked by pirates in recent months were the Ukrainian freighter *La Faina,* carrying 33 combat tanks and other weaponry, which was hijacked on the 25 September 2008, and the super tanker *Sirius Star,* taken in mid-November. The *Sirius Star,* a new ship worth approximately US$150 million, is the largest vessel ever taken by pirates and was carrying a cargo of crude oil with a value of US$100 million at the time of attack. While the *Sirius Star* has been released in January 2009 after a ransom of reportedly US$3 million was paid, the hijacking of the super tanker clearly demonstrated the capacity of the Somali sea-robbers to attack ships of any size. Furthermore, attacks in recent weeks, such as the hijacking of the tanker *Sea Princess II* on January 2, show that current efforts to combat piracy in the region are not sufficient to prevent major attacks – despite the involvement of naval forces from countries around the world.

Responses

The vast majority of pirate attacks around the world occur in national waters, and piracy is therefore often considered a problem that should be addressed by the state (or states) where attacks take place.

Somalia, however, does not at present have the capacity to secure its waters. The country has often been described as a failed state, and has had no effective government in place since 1991. Indeed, after the end of the Cold War, the country's central government collapsed and Somalia has been ruled by a succession of varying coalitions of politicians and local warlords. With weapons widely available, armed conflict and violence has been a constant component of 'politics' in Somalia. Famine and other natural and manmade disasters have been a further long-term burden for the country's population. International organisations, such as the UN, and various countries from within and beyond Africa have, throughout this period, been involved in Somali politics and conflicts and have provided limited humanitarian assistance.

Over the years, efforts were made by local and international governments and interest groups to stabilise the political situation in Somalia, with a succession of interim governments in place. In early 2006, with the fragile Transnational Federal Government (TFG) in power, fighting between the Union of Islamic Courts (UIC) and the US-supported Alliance for the Restoration of Peace and Counter Terrorism (ARPCT) erupted. The UIC gained control over Mogadishu and large parts of the south, and a ceasefire agreement was reached in June 2006. While the UIC held power, pirate attacks-which had been a concern in Somalia-ceased, as UIC fighters defeated armed groups involved in piracy and announced that pirates would be punished under Sharia law. However, the situation deteriorated again when US-backed Ethiopian troops entered the conflict in support of the TFG, gaining the upper hand in December. The defeat of the UIC did not, however, bring peace or stability, despite a UN Security Council-authorised African Union peacekeeping mission, starting in February 2007. In fact, the continued presence of Ethiopian troops fuelled further unrest in a country already plagued by internal unrest and clan-based and political rivalry.

In early 2008, the security situation deteriorated once again with escalating armed conflict between Islamist insurgents, Ethiopian troops and other factions in addition to US air strikes on Islamist bases as part of the war on terrorism. Today, the situation is still volatile, with the country's President Abdillahi Yusuf resigning in early January 2009. Furthermore, Ethiopia has begun to withdraw its troops from Somalia, having made little progress towards stabilising the country. Peacekeepers from the African Union will fill some of the vacated positions to avoid a power vacuum.

The long internal conflict in Somalia combined with natural disasters, such as water shortages and severe droughts, have left the country devastated. Hundreds of thousands of people have lost their lives and an estimated 2.5 million are in urgent need of assistance. Since January 2006 alone, an estimated 1.1 million people have been displaced in Somalia, with many seeking refuge in neighbouring countries.

Given the local political instability, authorities in Somalia are unable to secure shipping in their waters. Indeed, local authorities, as far as they exist and function, often have more pressing issues to address. Furthermore, anti-piracy measures that have worked (at least to some extent) in other places that have relied on government forces of countries in which attacks take place, are also difficult to implement in the Gulf of Aden and off the Horn of Africa. In the Malacca Straits, for example, increased cooperation between the littoral states has reduced the number of pirate attacks. Given the current political situation in Somalia, such cooperation would be difficult – even if the necessary equipment was available to government forces.

International Responses

With local forces failing to secure Somali waters and the frequency and scale of pirate attacks posing a serious threat to global shipping, international concern grew, triggering two different kinds of responses. First, ship and cargo owners hired private security companies (PSCs) from different parts of the world to secure vessels transiting the pirate-infested waters. Second, nations from around the world have sent warships to the area in addition to those international vessels already present. Warships from the US, Britain and other countries have been patrolling the waters off the Horn of Africa for the past eight years to combat terrorism. The additional vessels were sent in recent months mostly from those countries whose ships have been targeted by pirates. Among the warships and personnel patrolling the area have been ships from the US, Canada, Britain, Germany, France, Russia, Malaysia and India. Many of these vessels are part of missions sanctioned or organised by multilateral organisations. Both NATO and the EU have, for example, been involved in combating piracy in the Gulf of Aden, with the NATO mission ending in December 2008, when EU efforts began. Both organisations have sent warships to patrol pirate infested waters, placing particular emphasis on the protection of UN aid vessels.

This international presence has prevented a number of hijackings, with naval forces from a variety of countries having engaged in fire fights with pirates to foil attacks. British forces, for example, killed three pirates in a shootout in November 2008, while EU naval forces prevented an attack on a crude oil carrier on 2 January 2009. France, which maintains a naval base in Djibouti, has also conducted successful operations against the pirates. In April 2008, French Special Forces arrested six pirates responsible for an attack on a French cruise yacht with 30 people onboard. After the hostages were released the pirates were captured in a daring helicopter attack on a jeep carrying the perpetrators and part of the ransom money in the semiautonomous province of Puntland. In September, French Special Forces recaptured a hijacked yacht 300 miles off Somalia. The pirates had taken the yacht and a French couple hostage, demanding a ransom of Euro 1 million and the release of the pirates earlier arrested by French authorities. Soldiers from the underwater combat unit successfully boarded the yacht at sea, shot one pirate and arrested the other six perpetrators. Furthermore, every two weeks, the French Navy organises an escort for ships passing through the pirate infested waters. Merchant vessels in the area at the time can join the group free of charge.

The latest country to join the international effort to combat piracy off Somalia is China. It is the Chinese Navy's first operational deployment outside Asia, with the country sending two navy destroyers and a logistics ship to the pirate-infested waters in early January 2009. International efforts were further boosted that month with the creation of Combined Task Force 151, which became fully operational in mid January. The task force was set-up by the US 5th Fleet, based in Bahrain, and concentrates on combating piracy in and around the Gulf of Aden, the Arabian Sea, the Indian Ocean and the Red Sea. It was established in addition to Combined Task Force 150, which focuses on other security operations in the area, such as anti-smuggling. Head of the new anti-piracy task force is US Rear Admiral Terence McKnight of the US Navy. More than 20 nations have been invited to join the task force, among them Australia.

A Role for Australia?

Following the invitation to join the task-force Defence head Air Chief Marshal Angus Houston stated that the Department of Defence would consider deploying an Australian warship to the pirate infested waters off the Horn of Africa. Addressing sailors onboard the *HMAS Parramatta* in the Persian Gulf on 7 January 2009, he commented that different deployments of Australian warships are considered:

There is... the possibility of perhaps going further and getting involved in some of the important counter-piracy work that is coming on in the north Indian Ocean, (...) (We will be) looking at the options available and making recommendations to government on the best way to go. With Australia's role in guarding Iraq's offshore oil facilities coming to an end, a Defence spokeswoman confirmed that: "Defence is prepared for a range of contingencies, including the possibility of contributing to UN-sponsored anti-piracy operations off the Horn of Africa" but added: "No decision has been made and the deployment of Australian forces is a matter for government." Joining the debate, Defence Minister Joel Fitzgibbon voiced his concern about Somali piracy, but emphasised that he had not decided whether or not Australia should join the anti-piracy task force, despite the ADF's capacity to "respond to a range of global contingencies".

The suggestion to send an Australian warship has already found support in some quarters. Teresa Hatch, executive director of the Australian Shipowners Association, for example, believes that Australia should contribute to the anti-piracy task force, pointing to the importance of the Gulf of Aden to international trade. Anthony Bergin, Director of Research Programs for the Australian Strategic Policy Institute further stressed the importance of maritime trade for Australia, arguing that the protection of international shipping is of high strategic interest to Australia. Indeed, the Gulf of Aden is of vital strategic importance, with more than 20,000 vessels travelling between Europe and the Middle East, Asia and Australia passing through the waterway every year. Among them are Australian merchant vessels, ships with Australian crew onboard, as well as vessels carrying cargo to and from Australia. Australian cruise ships and those carrying Australian holidaymakers also sail through the Gulf of Aden and are targets for pirates. In fact, cruise ships from various countries have in the past months repeatedly come under attack, including those carrying Australian passengers. Participation in the international anti-piracy efforts would therefore help protect Australian nationals and assets as well as national strategic interests.

Australia's participation in the UN approved anti-piracy task force would also strengthen Australia's role in international peacekeeping and security missions and support efforts by the international community – in the form of multilateral institutions- to address pressing security threats. For example, sending an Australian warship to Somalia as part of the UN approved mission would lend additional strength and resources to the international task force. This may not only help reduce the number of pirate attacks in this area, but also strengthen the UN's role in addressing non-traditional security threats. Since the end of the Cold War, non-traditional security issues that in the past were not perceived as being part of the international security agenda have become increasingly important. Over the past two decades, it has also become more and more apparent that many non-traditional threats, such as terrorism, transnational crime or environmental degradation cannot be addressed by one state alone. In fact, the lines between national and international security have become blurred. In response, multi-lateral institutions, such as the UN or NATO, have become increasingly involved in addressing security threats, but have often encountered operational difficulties. Any support for UN approved missions, such as the deployment of a warship as part of the anti-piracy task force in Somalia, can contribute to the success of such operations and consequently strengthen the UN's position as an institution capable of addressing and managing responses to international security threats.

Sending an Australian warship to support the anti-piracy mission would further confirm Australia's willingness to be an international citizen as well as the Royal Australian Navy's (RAN) capacity to respond to security threats around the world. While the Australian naval forces have been involved in numerous international assignments, (for instance, deployment to the Persian Gulf since the early 1990s) this would be an opportunity to participate not only in another high profile mission, but one that is in some respects remarkable. What sets this mission apart from prior operations is the participation of countries that have not previously deployed vessels far from their homeports. Most notably, this "is the first extended transcontinental naval operation deployment" for India and China, demonstrating the new role and capabilities of these countries' rising navies. It is also the first deployment of Western European vessels under an integrated EU command.

Participation in this anti-piracy mission would therefore also offer the opportunity to work together with naval forces from these and a variety of other countries and strengthen Australia's strategic international partnerships. Indeed, Bergin regards the possible deployment of an Australian warship as a chance to cooperate with other naval forces, particularly China, stating that:

Working with the [People's Liberation Army] navy is very much in Australia's interests (...) This can act as a confidence-building measure and build trust between other navies. It can generate understanding of other countries' procedures and assist in overcoming cultural barriers to military co-operation.

Cooperation with Chinese forces is of particular importance because the recent build up of naval strength in China has been viewed with interest – or even concern – in Australia. However, Bergin also warns that the rules of engagement of the Australian navy in the anti-piracy operation have to be clarified before a warship is deployed. Other observers have been more critical and have pointed out that the deployment of an additional warship will not solve the piracy problem. In fact, there are many shortcomings and problems inherent in current international responses to piracy off the Horn of Africa.

Shortcomings and Problems of Responses

Despite some success stories, current international efforts are clearly not solving the piracy problem, with pirate attacks continuing to occur for a variety of reasons. There are, for example, problems and shortcomings in regard to the rules of engagement of international forces in the fight against Somali pirates. The role and involvement of the international navies is restricted by national and international laws. To facilitate the involvement of warships from outside Somalia, the UN Security Council adopted four resolutions in 2008. In June 2008, for example, the Security Council unanimously passed Resolution 1816 which:

Essentially, (...) authorizes countries to enter Somali territorial waters and use 'all necessary means to identify, deter, prevent and repress acts of piracy and armed robbery' by boarding, searching and seizing suspect vessels and arresting the perpetrators. The key conditions require states taking such action to cooperate with Somalia's interim government and to notify the UN Secretary General.

In accordance with this resolution, international warships are able to patrol the affected waters to prevent attacks, and have created a so-called Maritime Security Patrol Area, a safe corridor for ships to pass through the Gulf of Aden. Yet, even within this zone the operations of international forces are restricted. For example, as NATO spokesperson James Appathurai explained, NATO vessels were only allowed to patrol, deter and stop attacks but not to use force to recapture hijacked vessels or rescue crewmembers.

To overcome some of these problems and to extent the authority of international forces involved in the anti-piracy operation, Resolution 1851 was passed by the Security Council in December 2008. It allows international forces with authorisation from the Somali government to pursue pirates on land until December 2009. While this resolution increases the capacity of international forces to act against pirates, operations on land are difficult to carry out, as has been demonstrated by earlier involvement of international forces in the country. One example is the 1993 Battle of Mogadishu, in which two US helicopters were shot down and three others damaged while on a mission to capture members of a powerful clan. Hundreds of Somalis lost their lives in the operation which ultimately led to the withdrawal of US troops from the country. Today, the civil war continues, Somalia is awash with weapons, and the pirates have at least some local support. Locating and arresting pirates on land is consequently a difficult and dangerous task that has the potential to exacerbate conflict in Somalia. As in earlier cases, operations by foreign forces in Somalia could result in the death of civilians and strengthen local antipathies towards foreign countries that have become involved in their homeland. In a worst-case scenario, such operations could intensify armed conflict and generate consequences, such as a further increase in crime or politically motivated violence that induce more foreign military responses.

Furthermore, the prosecution of pirates, caught on land or at sea, remains a problem. Jurisdiction is often unclear and pirates captured by international forces have been released in many cases because the arresting forces believed they could not convict the perpetrators under their respective national laws. As one commander describes: "It's frustrating, (...) we catch them, confiscate their weapons, and then we let them go." Indeed, pirates can often only be brought to justice in a foreign country if the victims (the vessel or the sailors) are from the same country as the forces arresting the pirates. France, for example, brought pirates responsible for two attacks on French vessels to France for trial, while handing other arrested pirates over to Somali authorities. However, the Kenyan government has recently agreed to prosecute arrested sea-robbers, with Britain, for example, having handed captured pirates to Kenyan authorities to be tried.

International responses also had operational deficiencies. Most worrying was the sinking of an alleged pirate mother ship by the Indian navy in November 2008. Unfortunately the ship was not a pirate vessel but a Thai fishing trawler that had been taken over by pirates earlier in the day. The trawler had a tracking device onboard and the owner of the vessel had notified the International Maritime Bureau (IMB) shortly after the hijacking. The IMB had broadcast the news to the coalition of navies involved in securing the waters off the Horn of Africa. While the British Navy confirmed the hijacking and backed away from the vessel, the Indian navy approached the ship later in the day, asking it to stop. A fire-fight between the two vessels ensued and the fishing trawler burned and sank. While two speedboats carrying pirates escaped from the vessel, only one of the 16 crewmembers of the fishing vessel was found alive-after drifting in the ocean for five days.

Another shortcoming of the operation which further explains why piracy persists is that despite the increased number of international vessels and the establishment of the Maritime Security Patrol Area, the warships cannot cover the entire area in which pirates can operate. The attack on the *Sirius Star*, which took place hundreds of miles south of the Gulf of Aden, demonstrates that the pirates are flexible, and will attack in waters outside the area protected by foreign navies. Furthermore, even if allowed, recapturing hijacked vessels is not an easy task and such operations have so far only been successful when small vessels (predominantly yachts) were rescued. Operations to recapture merchant vessels, including tankers, are far more complicated. Indeed, not only has the life of the crew onboard to be considered but a failed attempt could damage the hijacked vessel, and could, for example, result in a major oil spill. There are other reasons why the presence and operations of the warships have not brought an end to piracy. The most important reason, however, is that these efforts only address the symptoms and not the root causes of the problem. Piracy in this region is a direct result of the conflict, poverty and instability of Somalia. The pirates can only operate so successfully because they find supporters and willing recruits among the impoverished local population. As a36-year-old mother of five in Haradhere, close to where the super tanker was held, explains: "Regardless of how the money is coming in, legally or illegally, I can say it started a life in our town. (...) Our children are not worrying about food now, and they go to Islamic schools in the morning and play soccer in the afternoon. They are happy." Other factors such as illegal fishing and the ineffectiveness and corruption of government officials also play a role, with some observers suggesting links between Somali pirates and warlords.

Conclusion: What is to be Done?

Addressing and responding to piracy off Somalia is a difficult task. Indeed, piracy in these waters is likely to remain a problem until the political and humanitarian situation in Somalia improves. This is clearly a long-term project, requiring serious and sustained efforts by local leaders and the international community. However, given the scale of pirate attacks in this area, short term measures are also needed to prevent individual attacks, secure shipping and ensure that the situation at sea does not deteriorate further.

Australian Naval Deployments

The current effort by international naval forces is the most important of these short term measures to date. Despite shortcomings, international forces have been able to prevent numerous attacks and have demonstrated that pirates in Somalia (and other places around the world) cannot act with impunity. The deployment of an Australian warship would add further strength to this operation. As part of this multinational task force Australia would also support government-led, as opposed to private, responses to security threats. As discussed earlier, a second short-term measure to address piracy is the employment of PSCs. However, this can be problematic because these companies often operate in a legal grey-zone as rules and regulations regarding their activities are not clearly defined in places such as Somalia.

Government forces such as the Australian navy, on the other hand, are required to act according to UN set regulations and guidelines. Furthermore, a successful UN approved operation would also reinforce the institution's role in addressing and managing responses to non-traditional security threats. Finally, the deployment of a warship to Somalia is also an opportunity to strengthen Australia's profile in international security operations and to cooperate with naval forces from different countries.

However, there are still serious shortcomings with the current international responses. 'Errors' such as the sinking of the Thai trawler need to be avoided and information exchange between countries and institutions involved enhanced. Further improvements also have to be made in terms of the rules of engagement of international forces in Somalia and steps have to be taken to create (international) judicial mechanisms to bring arrested pirates to justice. A first step to address these issues was the formation of the Contact Group on Piracy off the Coast of Somalia. The first meeting of the group was held on 14 January this year at the United Nations, with Australia part of the meeting.

Australian Interventions for Long Term Somalian Human Security

While the creation of this group is a positive development, it is important to keep in mind that patrolling pirate infested waters will not address the underlying root causes of modern day piracy, which include illegal and over-fishing, lax (international) maritime regulations, ineffective and corrupt government forces, armed conflict and widespread poverty. Addressing these root causes is not an easy task and combating piracy can consequently not be achieved by only those states in which pirate attacks actually occur. It is therefore important for Australia, and other countries already involved in combating Somali piracy, to understand that the current international patrols have to be seen as a first step in a longer process.

Indeed, without long-term measures that address the political and humanitarian situation in Somalia, the patrols of pirate-infested waters are of limited use because they will only foil individual attacks, and only within the area covered by patrol vessels. This will not discourage pirates from operating off Somalia as unprotected vessels still pass through the area and pirates can conduct attacks outside the waters patrolled by international warships. With these short-term measures not solving the problem, international warships will have to protect vessels passing through these pirate-infested waters for the foreseeable future – unless long-term measures are successfully implemented and supported. Countries such as Australia, Japan and South Korea that consider sending warships to Somalia therefore have to understand that deploying warships could be a long-term commitment that may make little difference – particularly in regard to human security – if the political and humanitarian situation in Somalia is not addressed.

Consequently, the deployment of an Australian warship would make little sense unless Australia and other countries also take an active role in ensuring that long-term measures are taken and successfully implemented. For example, Australia could support efforts and policies that politically stabilise Somalia and bring an end to internal armed conflict. It is important in this regard to consider the interests of the Somali population and to support political solutions that are feasible as well as acceptable to the people of Somalia. In the past, international involvement in Somalia often had competing agendas, with some operations seeming to be driven predominantly by foreign rather than local interests. One example is the 2006 ousting of the Islamist government in Somalia by US backed Ethiopian forces. The US supported the campaign because it believed that the Islamists sheltered al-Qaeda operatives. The result was the end of a period of relative calm in Somalia during which there have been no serious pirate attacks. However, the resignation of the Somali President, who was seen by many as an obstacle to the peace process, and the withdrawal of Ethiopian troops from Somalia this January, may present another chance to relaunch the peace process. Yet, stability can only be achieved in Somalia if room is found for the different Islamist factions and any alliance of local groups – including Islamists – can only be successful if their efforts are not undermined by foreign forces. As long as a new government does not oppress the local population, respects the internationally guaranteed rights of its people as well as the borders of neighbouring countries, international support should be given. Indeed, foreign interests – such as eliminating hide-outs for terrorists – may best be served by supporting political solutions that will bring a government to power that has the ability to control and stabilise the country.

Australia could also back other efforts that aim at de-escalating the conflict in Somalia, strengthen the Djibouti peace process and enforce UN embargoes – such as the Somalia arms embargo. It could, for instance, support efforts to bring the violent US anti-terrorism campaign inside Somalia to an end. The campaign which is part of the broader war on terrorism has included airstrikes to kill suspected terrorists hiding in Somalia. However, the airstrikes have killed and wounded many civilians and have "weakened Washington's credibility in the Horn of Africa and galvanised anti-American feeling among insurgents and the general populace, as well as undermined the international effort to mediate a peace process for Somalia." The Somalia Arms Embargo Monitoring Group has also declared that these airstrikes, as well as US training of Somaliland officers in Ethiopia without exemption from the Security Council committee, are a violation of the Somalia arms embargo. While substantial changes in US foreign policy, including an end to airstrikes in Somalia, were unlikely under the Bush administration, there is hope that such changes may be possible now.

Furthermore, Australia could increase its support to address the grave humanitarian situation in Somalia. Short term, as well as long-term, assistance, administered through multilateral institutions such as the UN as well as NGOs, is needed to provide help to those affected by civil unrest and/or natural disasters. Finally, Australia could reinforce its efforts to combat other root causes of piracy, such as over and illegal fishing. In fact, support from countries such as Australia is needed to protect the marine environment and stop large fishing vessels, often operating under Flags of Conveniences (FOCs), from engaging in illegal fishing activities in waters throughout the world. In this regard, it is of particular importance to support initiatives of international bodies that seek to combat illegal and over fishing and attempt to address the shortcomings in the FOC system, including the International Maritime Organisation, the Food and Agriculture Organisation of the United Nations and the International Transport Workers' Federation.

While it is certainly a daunting task to successfully implement these long-term measures, they would hopefully lead not only to the eradication of piracy in Somalia, but also to more stability and security for the local population.

Piracy and International Law

Background: The Piracy Problem in the Failure of International Response

A series of spectacular hijackings has propelled the problem of Somali piracy the forefront of international attention. In the last months of 2008, different groups of pirates took a Ukrainian ship carrying battle tanks and arms to Kenya; a Saudi tanker filled with crude oil; a chemical tanker, and numerous ships carrying international relief supplies. Pirates have also fired on an American-owned cruise liner.

These are just the more notorious incidents in a rapidly escalating epidemic. Though only receiving public attention in recent months, attacks on international shipping have been a problem since the Somali government collapsed in the 1990s. Pirate attacks began growing in frequency and sophistication in 2005. Last year (2008) saw attacks on international shipping increase more than three-fold increase over the previous year – more than 100 ships have been attacked. Moreover, the geographic scope of attacks has widened.

In response, men-of-war from a 20-nation alliance, initially lead by NATO, began patrolling the Gulf of Aden. This has recently been augmented by a European Union flotilla — the first naval force ever deployed by the erstwhile economic union, as well as the first Chinese naval force ever deployed to another region, and Russian and Indians warships. The amassed armada is formidable. As one officer wryly noted, in a recent pursuit of a pirated vessel, there was roughly one destroyer per pirate.

Yet the results are not encouraging. The most modern navies have been dispatched against small, disorganized posse of petty criminals equipped with RPG's, small arms, and small fishing boats. Thus far, international efforts have failed to solve a problem that threatens to scare shipping away from a waterway that carries nearly one third of the world's cargo. While the hijacking of the Ukrainian *Faina* was said to be a bridge too far for the pirates – a whole armada of high-tech warships set off in pursuit – even that ship appears to have been successfully ransomed off. Despite extraordinary international cooperation in interdiction, the International Maritime Bureau reports an "unprecedented rise in pirate attacks."

While coalition vessels have stopped and prevented many pirate attacks, they have not taken any pirates off the streets. Instead, they have released all the pirates they apprehend because of jurisdictional concerns. Diplomatic and military officials have suggested that part of the problem lies in a lack of an international law basis for prosecuting pirates. As the Chairman of the US Joint Chiefs of Staff put it, "One of the challenges that...you have in piracy clearly is, if you are intervening and you capture pirates, is there a path to prosecute them? That's something I think the international community has got to answer for the long run."

These concerns are unfounded and misleading. International law provides an ample basis for prosecuting pirates. To the extent nations suggest otherwise, they are deflecting attention from their political unwillingness to use the robust prosecutorial options international law clearly affords. Unfortunately, their lack of willingness is quite predictable. For one, prosecution has costs that nations are reluctant to bear. Moreover, recent developments in international law create serious obstacles to military or judicial action against pirates. It is understandable that the legal and practical difficulties of dealing with international criminals deter third parties nations from doing so. Yet this may bode ill for the ability of Western nations to deal, within the constraints of international law, with even more complex cases of cross-border attacks by irregular forces, such as terrorist and guerrilla groups.

Multiple Avenues of Jurisdiction over Pirates

Universal Jurisdiction in the law of nations

Contrary to the protestations of many Western nations, international law as it currently stands gives ample license for pursuing and prosecuting Somali pirates. While international law may be vague and uncertain in many areas, piracy is not one of them. One of the oldest, clearest and least controversial rules of international law is that any country can prosecute any pirates they happen to apprehend. For hundreds of years, the pirate has been treated as a *hostis humani generis* – any enemy of all mankind. Several factors underpin pirates' special jurisdictional status.

First, international shipping passes through certain defined routes, and it is on these maritime highways the pirate threat. Thus the pirates, who do not discriminate among targets based on nationality, literally endanger the trade of all countries. Their attacks raise the prices of commodities and maritime insurance, and thus of international commerce in general. As recent attacks have shown, a ship may be owned by the one nation flying the flag of the second, carrying cargo destined for five other countries, and crewed by nationals of still other states. Piracy implicates the interest of all these countries.

Equally important, part of the definition of pirates is they are not endorsed by their own country. Thus when some other state seeks to prosecute them, it is unlikely the home country will protest. Based on the example of piracy, modern human rights law has attempted to extend universal jurisdiction to crimes with political motivations and governmental perpetrators, such as war crimes and torture. This has often raised the hackles of the defendant's government, of which the defendant was often formerly or currently a part. Pirates, by contrast, are mere robbers, over whom no one would have much solicitude.

While the reasons that piracy has historically been a universal jurisdiction offense a complex, the law is simple. Any ship patrolling off the Horn of Africa can prosecute any pirates it captures, even if the pirate has never attacked that that countries' shipping.

The United Nations Convention on the Law of the Sea (UNCLOS)

While universal jurisdiction over piracy developed through unwritten customary international law, it has been codified in the United Nations Convention on the Law of the Sea, which has been ratified by almost every country in the world (with the significant exception of the United States). UNCLOS defines piracy more broadly than did the customary law of nations. The treaty includes "any illegal acts of violence or detention, or any act of depredation, committed for private ends." Thus assault and murder on the high seas would be universally punishable under UNCLOS, whereas under traditional international law such crimes ironically had to go unpunished.

In some ways, UNCLOS might make fighting piracy harder. Other provisions of the treaty extend nations' territorial seas 12 miles from the coast, a significant expansion of the traditional three-mile zone. Moreover, the treaty gives coastal states certain exclusive rights as far as 200 miles out. Because the international law of piracy only applies on the "high seas," UNCLOS has the unintended effect of reducing the area where piracy can be internationally policed. This presents an opportunity for pirates to take advantage of the territorial waters of weak or failed states. This can have an important effect in gulfs, straits, and archipelagoes, where international shipping must transit through or close to sovereign waters. This helps explain why the two leading piracy problem areas of the Gulf of Aden and the Straits of Malacca, both choke-points for international shipping.

However, in respect to Somali piracy, this weakness in the UNCLOS regime was remedied by the United Nations Security Council in June 2008. Resolution 1816 authorizes the international force patrolling the Gulf of Aden to "enter the territorial waters of Somalia for the purpose of repressing acts of piracy" and while there to "use...all necessary means to repress act of piracy." The resolution is an unprecedented grant of authority to interdict coastal piracy. Enacted pursuant to the Council's Chapter VII powers to respond to threats to international peace, it appears to authorize a wide range of military measures measures. By removing a classic legal refuge of pirates – minimally governed sovereign coasts – the resolution, whose 6-month mandate was extended for another year in December 2008, grants legal powers to fight piracy that are arguably even greater than existed under the classic law of nations. The Security Council's action, which essentially authorizes the use of military force in Somalia territory, was taken with the explicit consent of the country's nominal transition government. Nonetheless in the wake of the resolution, pirate attacks became, if anything, bolder. Significantly, pirates began seizing vessels further from the Somali coast than ever before. One might speculate that this is because the resolution equalized the legal status of coastal waters on the high seas, reducing pirates' incentives to focus their attacks on the former. Or it may simply be that a series of successful and dramatic captures emboldened the pirate to strike further out.

Universal Jurisdiction in Practice

While universal jurisdiction prosecutions for piracy are quite rare, they are not unheard of. The nations patrolling the Gulf of Aden know that any country can prosecute pirates. Indeed, on two occasions in recent years they have handed over Somalia Pirates to Kenya courts for prosecution. In December Britain signed a memorandum of understanding with Kenya providing that in the future it would transfer all pirate captures there for trial. In none of these cases does Kenya have any connection with the crime. Rather, Kenya appears to have been chosen for convenience. But the jurisdiction exercised by Kenya in these cases is shared with all the nations of the world who could just as readily prosecute if they had the will.

Some of the European countries that have expressed doubts about whether they can prosecute Somali pirates have been at the forefront of exercising universal jurisdiction over other offenses. Denmark, for example, has suggested that it cannot punish the pirates. Yet it was one of the first European countries to prosecute Serb officers for crimes committed against Bosnian Muslims in the Yugoslav Civil War. International law poses no bar to Danish or any other country's prosecution of pirates. Denmark and other NATO countries have demonstrated that prosecuting foreign war criminals using universal jurisdiction is not incompatible with their legal systems or values.

The Suppression of Unlawful Acts Treaty

An additional source of jurisdictional power, which overlaps partially with the piracy norm, is the Convention for the Suppression of Unlawful Acts Against the Safety of Maritime

Navigation (SUA), enacted in response to the Achille Lauro hijacking. The treaty allows any signatory to prosecute anyone who "seizes or exercises control over ship by force or threat of force or any other form of intimidation." Jurisdiction is even more readily available for countries whose nationals are on the pirated vessel.

The Security Council's resolutions on Somali piracy have stressed the availability of jurisdiction under the convention. While SUA has 149 signatories, whose merchant marines comprise 92.75% of the world's shipping tonnage, prosecutions under the treaty have apparently been limited to a single case: obscure US case involving a deranged ship's cook who commandeered a fishing trawler. Indeed, the reluctance to use SUA as a basis for jurisdiction may be due in part to any lack of guidance about the precise applicability of the treaty.

Obstacles

Given the ample basis in international law for acting against pirates, hesitancy international leaders is on its face puzzling. However, a variety of second-order legal problems, as well as practical constraints and considerations of national self-interest, make piracy enforcement and prosecution difficult. For while international law authorizes indeed encourages the prosecution of international crimes, many particular rules actually create the opposite incentive.

War against Pirates Abolished

In the classic law of nations, pirates were simultaneously criminals and military enemies. This meant that they had all the disabilities of both classes. They could be captured and tried – unlike prisoners of war. If encountered on the high seas, the pirate could be attacked and killed. Moreover, international law recognized that returning captured pirates to a port for trial could be extremely burdensome for the capturing country. Thus summary shipboard proceedings were permitted. In short, pirates had a status much like unlawful combatants: they could be dealt with either militarily or criminally depending on the convenience of the capturing nation. Under modern international law, the situation is reversed. Pirates are not regarded as belligerents under the laws of war, and certainly not as illegal combatants. Furthermore, a central provision of the Law of the Sea convention reserves the high seas for "peaceful purposes." Except in situations of immediate self-defence, naval forces cannot make war on pirates, but rather must seek to apprehend them.

At the same time, given that pirates would be detained by armed forces and held, at least initially aboard naval ships, there would be strong pressure to give them all the protections of prisoners of war, at least presumptively. Indeed, in the theory of the Geneva Conventions widely held in the international legal community the campaign against the pirates, to the extent it takes place in Somali waters, could potentially qualify as an "armed conflict not of an international character." This would entitle pirates to the protections of common article 3 of the Geneva Conventions. In short, pirates today would have the protections of criminal defendants, as well as some of those of enemy prisoners, without the disabilities of either.

Indeed, given that pirates are civilians on board civilian vessels, there is little the patrolling navies can do until the suspected pirates attempt to board another ship. There is no international crime in being on a Somali fishing vessel, even if armed with AK-47s and rocket propelled grenades. If pirates are not stopped in the narrow window when they speed towards and boarded a vessel, anti-piracy enforcement turns into a hostage situation. Once the pirates have control of a vessel, there is overwhelming pressure to negotiate with them. Thus while the United States Navy came across a disabled motorboat bristling with armed suspected pirates, it gave them a lift rather than arrest them.

Forcefully dealing with pirates would create a serious risk of being accused of itself violating international humanitarian law. Simply attacking pirate vessel on the high seas would open nations to accusations of "targeted killing" of protected targets – similar to the kind of accusations that has plagued Israel in its campaign against terrorists. Moreover, in some cases, the accusations might not be baseless. This is vividly illustrated by one of the major armed actions of the anti-piracy flotilla. As the piracy problem continued to grow in November 2008, India reported that one of its frigates had sunk a pirate mothership in the Gulf of Aden. It soon emerged that the vessel was in fact a Thai fishing trawler that had been captured by pirates; only one of the 16 innocent crewmembers survived.

In another major December development, the Security Council passed a resolution sponsored by the United States allowing the use of military force on Pirates even on dry land in Somalia. For one year period, S.C. Res. 1851 authorizes nations to "undertake all necessary measures that are appropriate in Somalia, for the purpose of suppressing acts of piracy and armed robbery at sea." As another chapter 7 measure, it would allow for missile strikes another military assaults on pirate hideouts. This gives the coalition navies a tool the absence of which frustrated pirate hunters for centuries. This resolution makes even further inroads on the territorial sovereignty of Somalia. The proposal suggests that the United States at least no longer regards the purely criminal model of piracy as satisfactory and is seeking, at least in the context of Somalia, to return to the "war" model as well.

However, the resolution was immediately criticized – including by an American admiral in he Gulf of Aden – as likely to cause significant civilian casualties due to the difficulty of distinguishing pirates from anyone else. Such considerations have no doubt also dampened the international naval effort. In recent years international lawyers and NGOs have increasingly scrutinized tactical military decisions for proportionality and other indicia of legality. In wars against non-uniformed combatants and terrorist groups, national forces have been accused of violating humanitarian law when they did not successfully distinguish between combatants and civilians in an environment where the former freely commingle with the latter. In this environment, attempting to enforce international criminal law creates the danger of violating or being accused of violating international humanitarian law.

The same problem will now face national navies trying to deal with piracy. Indeed, the problem is more severe as the very notion of universal jurisdiction makes clear that the enforcing nation is not defending itself or its vessels. Moreover, even if the likelihood of actually violating humanitarian law is small, the potential perception that a nation kills civilians may be more important than an actual legal conclusion. And the small likelihood of crossing the line could be enough to deter nations that do not have any direct stake in the matter.

Difficulties in Prosecution

International law authorizes and encourages international prosecution of pirates. However, several separate developments in international law tend to make the detention and prosecution difficult and potentially embarrassing for the forum state. Interestingly, many of the difficulties closely parallel those encountered by the United States in its effort to prosecute foreigners captured abroad and suspected of involvement with terrorist groups.

Procedural Rights

Universal jurisdiction only applies to pirates. Based on past experience, captured Somalis are likely to insist in court that they are not pirates, but rather simple fishermen erroneously seized by a foreign Navy. Moreover, most Somali pirates are, in fact, fishermen; piracy is not a full-time job. Indeed, given that error is not out of the question, such challenges must be taken seriously, but greatly impede prosecution and detention. Numerous foreign nationals captured by the U.S. in Afghanistan claimed that they were not members of the Taliban or Al Qaeda, but rather foreign tourists, aid workers, or other innocent bystanders who happen to find themselves in Afghanistan. Moreover, U.S. courts have held that because the power to detain depends on the foreigner's status as a combatant, detainees can appear before tribunals to challenge the factual basis for being classified as a combatant even before a full trial for their alleged crimes. Establishing the very identity of captured individuals will be difficult, as they are unlikely to carry identification or other indicators of nationality.

Every aspect of the prosecution would raise serious logistical problems. Providing counsel and translation services into the defendants' native language would be extraordinarily difficult, even in neighboring Kenya. Transporting prosecution witnesses and evidence to a foreign court would be burdensome. Naval officers in active service would be called upon to testify. Identification by victims would be made even more problematic by the fact that multinational crews of foreign-flagged vessels would have returned to their homes around the world. Defence witnesses would present an even greater problem. Any provision for the defendants to be able to present evidence in their favour would require ongoing access to Somalia and Somali waters.

In the two cases brought in Kenyan courts at the behest of Western powers so far, the defendants claimed alleged torture and denial of religious privileges by their captors. Given that the Somalis are Muslim, the last charge could potentially be seized upon by Islamic countries to incite anti-western sentiment along the lines of incident in Guantánamo where a Koran was allegedly "disrespected." Indeed, it appears the pirates have read the Guantanamo detainees' playbook.

Other human rights rules of Western countries discourage arresting pirates in the first place, and certainly militate against trying them in Europe. Britain had reportedly affirmatively instructed its ships not to capture pirates they may come across. The Foreign Office feared the Somali pirates could claim asylum under European union human rights law once on board a British vessel, and certainly if brought to Britain for trial. Because of the routine nonexistence and brutality of the Somali government, the pirates could quite plausibly claim they would be subject to unfair trials, torture and extrajudicial killing if they are repatriated to their home country. Thus the pirates might be able to stay in Britain indefinitely if acquitted or once released. Obtaining European residency would doubtless be seen as a benefit to many Somalis, and the perceived reward for piracy might only increase the piracy problem.

In the face of such problems, it's not surprising that Britain in December took the unusual step of signing a memorandum of understanding with Kenya creating a framework whereby pirates captured by the former would be tried by the latter. For Britain this has numerous advantages. It escapes the embarrassment of having to release capture pirates, as Denmark has repeatedly done. In Kenya, the pirates do not benefit from European asylum law or the other expensive protections of the European Convention on Human Rights. The proceedings are relatively speedy and informal, with accusations of torture brushed away by the judge with the observation that the defendants don't appear to be bleeding. It is less obvious what Mombasa benefits from this. One might speculate that Britain has promised some sort of foreign aid or other benefits in exchange for the prosecutorial assistance.

The Free-rider Problem

Finally, there is the problem of incentives to prosecute. Policing against piracy faces a significant free rider problem. The nation taking action, either by apprehending or trying pirates, bears all the costs itself. Yet it enjoys just a small share of the global benefit of lower shipping and commodity costs. Nor does the deployment of a multinational force significantly reduce this problem. Each nation chooses whether to prosecute the pirates it apprehends. Indeed, the presence of warships from many nations in the Gulf of Aden actually increases the free rider problem.

Moreover, in individual cases, a nation will know whether a ship being attacked is of particular or direct interest to it. Nations will accept the high costs of prosecution only when their interests are directly affected, such as when its vessels or nationals are targeted by pirates. Thus the two instances where Somalia pirates have been taken to Europe for trial, neither involve UJ. In one case, pirates seized a French yacht. Paris responded with extraordinary vigor, sending commandos into the mainland of Somalia to liberate the passengers and crew, who were being held hostage. France identified the pirates, and brought them back to Paris for trial. Similarly, Holland plans to try pirates that attacked a Dutch vessel. The suspects are being held by Denmark, which says it will not exercise UJ over piracy. Thus, while pirates legally are called enemies of all mankind, in practice states act as if they are enemies of the countries they attack.

The abject failure of the international response to piracy in the Gulf of Aden is a cautionary tale about the limits of international law and the genuine lack of interest nations have in enforcing international law norms when doing so is costly for them. At first glance, the piracy problem has all the hallmarks for a successful international legal response. The scale of the problem is truly international — any country with commercial interests is adversely affected by the endangerment of shipping in an area through which much of the world's maritime traffic travels. (The seizure of Saudi oil tanker in November sent crude oil prices up by a dollar a barrel). Moreover, those responsible for the problem are a few thousand disorganized, illiterate, poorly armed brigands. The only state whose sovereign interests are involved – Somalia – has made it clear that it has no intention of intervening on behalf of the pirates. The most august institutions of international law making have focused their attention and authority on the problem. The Security Council has passed five resolutions dealing with Somali piracy in 2008 alone, all under the solemn Chapter 7 authority to permit the use of military force. No issue – not even the Israel/Arab question – has been the subject of as many resolutions last year. The piracy problem has enjoyed the advocacy of important international organizations such as the International Maritime Organization. Finally, piracy is the paradigm crime for which international law authorizes enforcement and punishment. Moreover, in a genuine act of international cooperation the major powers have committed significant naval forces to deal with the problem.

Thus the spectacle of the international community wringing its hands, looking for a legal solution to the piracy problem, can be understood only as unwillingness to use the international legal tools due to the costs involved. This is ironic given, under the slogan "an end to impunity," universal jurisdiction has in recent years been presented as a solution to any crime, regardless of the nationality of the alleged perpetrator or victim and regardless of the connection of the state exercising its jurisdiction in the manner. European counties have launched prosecutions against political and military leaders of the various countries. In all of these cases, the inevitable political overtones always call into question the motives for the prosecution.

Yet these countries cannot bring themselves to prosecute a single case of piracy universal jurisdiction, even though it means allowing a major ongoing international problem go unchecked. Indeed, the United Kingdom's decision to render all pirates it seizes to Kenya underscores the strength and weaknesses of the current international regime.

Many of the difficulties involved in apprehending and prosecuting pirates parallel those involved in making cases against detainees held by the United States in Guantánamo Bay. The possibility of a criminal model for dealing with international terrorists seized abroad has been much discussed. The apparent rejection of such a model by all Western nations as applied to piracy suggests there is little reason for optimism when it comes to the more politically fraught context of terrorism.

Cruise Ships & Pirates: How Safe are You?

As the whole world knows, on Sunday, Nov. 31 Oceania Cruises' 684-passenger Nautica was fired upon by pirates in the Gulf of Aden. When officers spotted the small boat, Nautica increased speed and easily left the pirates in the dust — er, in the water. From start to finish, the incident lasted five minutes, hardly a disruption in the voyage.

But this is the third time cruise ships have been approached by Somali pirates in the Gulf of Aden so the situation bears exploring. First of all, cruise ships either repositioning (the Mediterranean to Asia) or on a world cruise cross the Suez Canal because the only alternative would be to sail around Africa, a continent loaded with very boring ports.

Cruise lines have a firm policy of never discussing security. Those East Asian men checking you on and off the ship are typically Nepalese Ghurkas, considered to be the finest soldiers on earth. You may spot an Israeli, also top notch. Prior to 9/11, this was the extent of a cruise ship's security force. And while European warships patrol the gulf, the Somalis have been very successful at stoping ships and boarding them.

According to the web site ShipTalk, there have been around 95 pirate attacks in 2008, with 40 freighters held for ransom.

By the way, when you think about Somali pirates trying to capture a cruise ship, do not envision Johnny Depp. Somalia has no government, its population is starving, so some citizens try to make a living by capturing a ship — cargo or passenger, it doesn't matter — to hold for ransom. The pirates, who apparently feel misunderstood, recently agreed to be interviewed by the New York Times saying they only want food and money. Shiptalk reports that companies around the world are increasing private security firms to accompany vessels through this area. Nick Davis, chief executive of the Anti-Piracy Maritime Security Solutions, says his company has 10 teams on board ship in the region. According to Davis, when pirates approached a ship with his men aboard, they abandoned the vessel upon discovering that there were professionals aboard. Personally I won't be surprised if cruise ships start carrying weapons for self-defence when cruising in the Gulf of Aden.

Seabourn Cruises' ships have been approached twice by pirates. During the first incident in 2005, officers turned a Sonic Gun on the pirates. Created by the U.S. Defence Depart, the device emits an hellishly loud sound that repels pirates. While the government is sharing its technology I have no way of knowing which cruise lines use it.

So, should you be concerned enough to avoid cruising in this area of the world? It wouldn't stop me. You see there's a big difference between a freighter carry a crew of 10 who know how to tie ropes and a cruise ship with hundreds of staff, including the best soldiers in the world.

Cruise Ships as Soft Targets

A second important point to consider about the GAO report is that it is exclusively focused on the security of cruise ships operating in U.S. territorial waters. When one recognizes that cruise ships traditionally operate in foreign waters and that cruise lines take their fleets of ships to foreign ports of call, the following finding in the GAO study should give passengers and crew, as well as cruise line security planners, a reason to be concerned. "According to an official of the Cruise Lines International Association, of the cruise lines included in our (GAO) site visits, only one had a vessel registered in the United States. Hence, although they carry large numbers of U.S. passengers, the vast majority of cruise-line operated vessels generally come under U.S. authority only when they enter waters over which the United States has jurisdiction" (emphasis added).

The implications of this fact is obvious; at the very least cruise ships are a greater risk from terrorists and pirates when operating outside of the territorial waters of the United States and at worst, the United States government and its vast law enforcement resources and even its military have limited or no jurisdictional authority or mandate to protect cruise ships on the high seas. Even if "no credible threats" had been received by U.S. intelligence agencies, it is safe to assume that cruise ships as the government report makes clear, remain highly attractive targets to terrorists representing high prestige and symbolic targets to well known international terrorists groups such as al-Qaeda.

Maritime security is perceived in recent years to have been boosted through the implementation in July 2004 of the International ship and Port Security (ISPS) Codes. Most of these security efforts are aimed at preventing terrorists from gaining access to ports and/or ships and inserting weapons of mass destruction (WMD) into the maritime domain. Far from port however, cruise ships on the high seas face peril from new pirate tactics that mimic terrorist attacks and from terrorists who seek to take advantage of maritime "soft targets." In "Cruising for Trouble," I examined these risks as well as the inherent risks to cruise-ship passengers onboard the new "mega" cruise ships coming into service in the last decade. These risks include those assumed by the ship and its passengers when visiting foreign ports. My experience as the Director of Security for Princess Cruises was used as a basis for evaluating cruise ship security. In Cruising for Trouble, I was critical of the security preparations of the cruise industry on many levels.

The industry has experienced a remarkable expansion of its fleets with the addition of ultra-large luxury liners that cater to every pocketbook and vacation experience. Cruise ships continue to grow in size and develop new marketing strategies to tap into the huge North American market where according to the cruise line's own statistics; only 17 percent of the U.S. population has ever taken a cruise. The cruise industry has recovered relatively quickly from two wars in the Middle East, a SARS epidemic, a devastating tsunami, several cruise ship fires, collisions and sinking's and even (at last count), three pirate attacks. These and other cruise ship stories can be found detailed in Cruising for Trouble. The industry however has also been a victim of its own phenomenal growth. Cruise ship incidents involving mysterious passenger disappearances, (man-overboard), underage drinking, drug use and drug smuggling as well as sexual and physical assaults have continued to occur at an alarming pace. It is only natural to assume that such incidents would follow when increased passenger loads on cruise ships in previous years averaged 2,000 passengers but is now averaging 3,500 with even bigger ships coming online. Passenger loads on the new Royal Caribbean ship "Oasis of the Seas" for example launched in November, 2009, approach nearly 6,500 (excluding crew).

Notwithstanding the phenomenal potential for safety and criminal incidents occurring during passenger excursions ashore, the reasons for such security occurrences onboard cruise ships is easily understandable. Cruise lines hire personnel to act as "security" on their ships and provide no real law enforcement response to the sexual assaults, bar-room brawls or thefts of personal property. Worse, there is no international standard for the prevention of crime onboard ships or for professionally investigating crimes committed at sea. The disappearance of George Allen Smith IV off Royal Caribbean's the Brilliance of the Sea in the Aegean Sea in July 2005 is a perfect example. The criticism of the FBI to fully investigate the crime and the professionalism of the Turkish investigation as well as the actions of the ship all combined to present a questionable and haphazard response to the loss of a man's life at sea which went so far as to hint at a cruise-line cover-up. The attitudes of the cruise line executives who run these fleets are no less encouraging to the prospective cruise ship passenger. The former CEO of Carnival Cruise Line who referred to the disappearance of George Smith as a "non-event" seems to trivialize such occurrences and does not seem to be in sync with the safety of lives at sea.

In June 2010, the passage of the 2009 Cruise Vessel Safety and Security Act sponsored by Sen. John Kerry (D-MA) and Rep. Doris Matsui (D-CA) with collaboration between the International Cruise Victims Organization (Internationalcruisevictims.org) and the Cruise Lines International Association (CLIA.org) brings some needed reforms to the cruise industry with regards to passenger safety, however further international efforts will be needed to make cruising a truly safe environment for all passengers and crew wherever cruise ships set sail. Currently, the U.S. law does not apply to cruise ships operating outside of U.S. jurisdiction. Efforts underway by the International Cruise Victim's Organization in Europe seek to make similar security standards and regulation applicable to cruise ships operating in Europe and Australia. Real progress in reforming the current cruise ship security model however will only be addressed when the international maritime community adopts the standard for placing recognized law enforcement officials on cruise ships. This is an idea that the cruise lines will vehemently oppose. With the sea-going public's safety in the balance on these new mega-cruise ships, and as mysterious disappearances and sexual assaults continue to increase; this standard will ultimately become a reality. The alternative is that the cruise industry's burgeoning fleets of mega-cruise ships will in turn, become ghost ships.

The attack on Seabourn Spirit may have highlighted security issues but concern for safety dates back to Titanic.

By any measure it was one of the more bizarre incidents in recent cruise history. On November 5, 2005, a group of "pirates" tried to attack Seabourn Spirit. The ship was making its way southward 160 kilometres off the coast of Somalia, when bandits in two speedboats approached the vessel brandishing a rocket-propelled grenade launcher.

Quick thinking and skillful action by the captain allowed Seabourn Spirit to evade the pirates. Employing the ship's superior speed, he was then able to outrun the attackers. What was even more astonishing was that the crew employed a "sonic weapon" to disable the pirates by blasting them with high-pitched noise, according to press reports. After the event, the Washington-based International Council of Cruise Lines (ICCL) issued a statement commending the captain and the crew. "Due to the highly trained security staff on board," the council said, "the ship was able to immediately initiate measures to safeguard the passengers and vessel." The council went on to emphasise the unprecedented nature of the attack. "This is the only known piracy attack on a cruise ship."

While this remarkable incident may have been unique, concern for passenger safety goes all the way back to Titanic, when the seafaring nations of the world joined together to create the

International Convention for the Safety of Life at Sea (SOLAS). Administered by the International Maritime Organization (IMO), an agency of the United Nations, SOLAS has been called "the most important of all treaties dealing with maritime safety." It was first adopted in 1914 and has been updated and revised many times since.

Indeed, SOLAS has some very visible components that all cruise passengers can see for themselves. For example, it was this convention that, in October 1997, mandated all major cruise ships to install smoke detectors, sprinklers and low-level lighting-similar to the coloured lights along the floor in airplanes which lead passengers to safety in the event of a power failure.

SOLAS is also the reason why every cruise ship conducts a lifeboat drill. Under the terms of the convention, these drills must take place within 24 hours of the ship setting sail. The drills are mandatory for passengers and crew alike. Everyone must be present and attendance is taken. All other non-essential ship operations cease during this time-no food is served and the bars are closed. Once passengers are familiarised with the ship's security procedures-how to don a life vest and where their muster station is located-they are dismissed and shipboard life returns to normal.

Ever since the events of September 11, 2001, cruise lines have upped their security measures to prevent unwanted elements from boarding the ship in the guise of a passenger. For example, cruise lines now ask passengers to pre-register their passports and other personal information online before arriving at the pier. This is to allow for a manifest to be submitted to local government authorities before the ship embarks. It's the same policy was put into place regarding international flights after the World Trade Centre attacks.

Cruise lines are generally tight-lipped about the specifics of their security policies. After the attack on Seabourn Spirit, the line would neither confirm nor deny the presence of a LRAD (Long-Range Acoustic Device) aboard its ship. "As a matter of policy, we will not discuss security equipment nor tactics," says Bruce Good, director of public relations for Seabourn Cruise Line. Carnival Cruise Lines is equally mum when asked about its anti-terrorist procedures. "We are not permitted to discuss specific elements of our security program so as not to undermine those efforts," explains Jennifer de la Cruz, director of public relations.

However, in a recent hearing before the United States Congress, Gregory Purdy, Royal Caribbean International's corporate director of safety, security and environment, gave a rare public glimpse inside cruise line security procedures. Because the major cruise lines are headquartered in North America, they are subject to a certain degree of oversight by the United States government. For example, the US Centres for Disease Control in Atlanta conducts the Vessel Sanitation Program, an arrangement where all cruise ships that call on US ports are subject to voluntary inspections for food handling and other hygienic procedures.

According to a transcript of his testimony before the House Subcommittee on National Security, Emerging Threats and International Relations, provided by Royal Caribbean, Purdy outlined a number of steps the cruise industry takes to secure its vessels.

"Every one of our ships has a security officer, who is a trained military veteran or career security specialist. Their sole responsibility is to ensure the safety of our guests and crew. These security officers oversee teams of trained security staff who are responsible for security procedures aboard our ships 24 hours a day. Surveillance cameras throughout our ships assist these efforts," he said. "Our onboard security operations also include trained divers on each ship," Purdy continued. "These divers can search ship hulls and ports at which we call, and work in coordination with various law enforcement agencies."

While the cruise lines are prepared for the unlikely event of a terrorist assault, the most common and more likely crimes you're bound to face aboard ship are the same as at home-a petty theft or minor assault. "Crime can touch any one of us in virtually any setting," Purdy told the Congress. "It is impossible to be immune to crime." He further stressed that: "One crime is too many. Our company maintains a strict zero-tolerance policy on crime," Purdy went on to conclude that cruise ships are one of the safest ways to travel and crime aboard cruise ships is extremely rare. Rest assured that if you have a problem aboard ship, the staff are there to help. Report any suspected theft or other crime to the front desk, advises Erik Elvejord, director of public relations for Holland America Line. If you're not satisfied with their response, speak to the ship's top security officer. "Then you're talking to the chief of police," Elvejord says. If your complaint involves a lost or stolen item, the security team will investigate. "We can do searches on a ship," he notes. "If a cabin steward is in possession of the stolen property, they will be fired on the spot."

While most of the crime on ships involves theft, there have been times when passengers have become unruly. In such cases the offending passenger will be escorted to his cabin. If necessary, security guards will be placed outside the cabin and the passenger will be confined under "house arrest", Elvejord explains.

Elvejord points out that passengers travelling with valuables should take steps to ensure their security. For starters, "Use the safe in your cabin," he says. "There's no reason to be walking around the ship with large amounts of cash. Also, if you travel with expensive jewellery, provide the ship with documentation. It can be very helpful should something missing turn up," Elvejord concludes.

Safe and Secure

After the events of September 11, 2001, the cruise industry put into place a variety of procedures intended to foil a terrorist attack. Most of these policies are still in effect today and include:

- A 90-metre security perimeter around ships while in port Airport-style x-ray screening of all carry-on baggage
- Pre-and post-cruise bomb searches aboard ships Background checks for all crew
- X-ray screening of all provisions
- Guest manifests checked against terrorist watchlists.

Pirates, how do Cruise Lines Deal with the Threat

Pirates are a worrying problem with cruise lines. It has been reported on several occasions that pirates have boarded ships and sailing vessels in certain water's in the world.

These pirates are vicious people who are usually heavily armed and will go to any lengths to attack smaller vessels.

Here is part of a document which passengers on QM2 recieved on their world voyage:

Between the 30th March and the 6th April, from Cochin to Dubai and onto the Suez canal, we will be transiting an area well documented in the world's media for piracy, in particular the Gulf of Aden, which we will transit on the 5th and 6th April.

Onboard Queen Mary 2, we have put in place a security plan in order to protect the ship.

As part of this plan, I would like to request your assistance and cooperation with the following:

Deck 7 external promenade deck and Deck 6 aft will be closed on both sides at night and no Guests or crew will be allowed access for the duration of the transit.

During hours of darkness, only essential open deck lights will be on as it is important to reduce the ship's external lighting. With this in mind I request that you turn off your cabin lights when not required and close the curtains of your window or balcony.

In order to familiarize you with these additional measures prior to transiting the Gulf Of Aden, there will be a Guest and Crew drill at 10.30 on the morning of Monday 4th April. This Drill will be conducted in the interests of your personal safety and security and I respectfully request your full participation in this important exercise.

In the event of pirate activity, announcements will be made from the Bridge and the initial actions required by our guests will be as follows: Guests are to proceed to their staterooms Guests with a window or balcony should proceed to the corridor outside their stateroom and sit on either a chair or on the floor and close their stateroom door. If you are in an inboard stateroom please remain inside your stateroom. Your Stateroom Steward will do a check of their section area to ensure everyone is accounted for. During this time Guests are not to remain on any open decks or near any external windows.

I must emphasise that the safety and the security of every individual on board is my highest priority and the implementation of this procedure is an important part of the ship's security plan.

Now this is excellent safety advice from Cunard and they seem to have thought of the scenario in case it did happen. Safety drills are an important aspect of cruising and they can be taken light hearted by people who have cruised many times.

Health Safety and Security

Should Travel Agents Be Liable For Falsely Representing That Cruises Are Safe For Kids?

Today I read a travel agency's twitter feed "Safety for Children on Cruise Ships." I am interested in this issue after representing 60 clients sexually assaulted on cruise ships, including children.

I was hoping to see some good safety tips in the article to keep children safe. Boy was I disappointed. Not only were there no useful safety tips but the travel agents made the following comment:

... most parents also like the fact that a cruise also allows you to spend some time apart from your kids, knowing that they are well taken care of.

Royal Caribbean, Disney, Celebrity and many other cruise lines have special programs for children, supervised by crew members who are specially qualified and trained to work with kids.

Unfortunately, this is simply not true.

We have represented many families whose children have been molested by youth counsellors on cruise ships. Other children have been victimized by crew members or even other passengers when they were suppose to be under the supervision of the counsellors. Others have been taken into cabins by older teenagers and men and raped when they were trying to walk back to their cabins after the kid's activities ended.

Another dangerous situation is when parents leave their kids in the cabin to see a show, or they want to stay up late in the casino or disco. A cabin attendant can open the cabin door with his pass key when the parents are away. As I have said in a prior article, leaving a 12 year old boy or his 14 year old sister alone in a cabin is a bad idea.

Cruise lines suggest that the crew members are carefully screened before they are hired. Nonsense.

The cruise line themselves do not perform background checks at all. They rely on the hiring agents around the world to perform all of the "screening." But what type of background check can be performed in countries like Nicaragua, Trinidad, or India? These places do not have computerized databases for drivers licenses or social security numbers much less a database for criminals or a tracking system for sexual predators. Many of the hiring agencies who are suppose to be conducting background checks are paid commissions when they place employees on cruise ships. There is a disincentive to keep the bad apples off of the cruise ships.

Many travel agents shy away from unpleasant facts. But it is another thing to affirmatively represent that children are safe and crew members are "specially qualified and trained."

Exactly what "special qualifications" does the crew member you entrust your child have? Why don't you ask the crew member before you drop your child off in the "kid's club." Don't be surprised when they look at you with a puzzled face.

If you are a travel agent reading this-do you think that you face liability if you make a representation that kids are safe to a client and a crew member molests your client's child during the cruise you just sold? I know the answer. Do you?

Before you make a false statement in the hope of making a sale, take a moment and read the following articles:

3

Disaster Management Techniques

The Titanic Disaster

An Enduring Example of Money Management vs. Risk Management

Most of the discussion of the accident revolves around specific problems. There was the lack of sufficient lifeboats (enough for at most 1200 on a ship carrying 2200). There was the steaming ahead at full-speed despite various warnings about the ice-field. There was the lack of binoculars for the lookout. There were the poor procedures with the new invention, the wireless (not all warnings sent to the ship reached the bridge, and a nearby ship, the operator abed, missed Titanic's SOS). Very recently, from recovered wreckage, "Popular Science" claimed the hull was particularly brittle even for the metallurgy of the time. (A claim now debunked.) Each has at one time or another been put forward as "THE reason the Titanic sank".

What gets far less comment is that most of the problems all came from a larger, systemic problem: the owners and operators of steamships had for five decades taken larger and larger risks to save money-risks to which they had methodically blinded themselves. The Titanic disaster suddenly ripped away the blindfolds and changed dozens of attitudes, practices, and standards almost literally overnight. The perception persists that the Titanic was, if obviously not "unsinkable" (though the White Star line actually never used that word in advertising), then very safe, as safe as the art could build her. That, despite various errors, the accident was mostly enormous bad luck. Nothing could be further from the truth. It was amazing good luck that there had been no similar accidents years earlier. For over 50 years, safety standards had been steadily deteriorating in various ways-almost always because of pressures to be "competitive".

Walter Lord, author of the classic "A Night to Remember", describes the process vividly in his 1986 sequel, "The Night Lives On". He compares the ships of Titanic's day to the first great liner, the "Great Eastern", built in 1858. She was designed by I.K. Brunel, England's most celebrated engineer, who got every feature he wanted. The Great Eastern was not the most profitable ship, but she was a triumph of safety. She had an entire inner hull two feet inside the outer. Inside that, the ship was divided by 15 transverse bulkheads, and one lengthwise into 32 compartments. Watertight lower decks further divided those.

The decades passed, and dozens, then hundreds of liners were built. Competitive pressures between some 11 lines were fierce. As Walter Lord relates:

"But the engineers did not have the last word for very long...the perfect ship was no longer the vessel that best expressed the art of the shipbuilder. It was the ship that made the most money."

"Passengers demanded attention; stewards could serve them more easily if doors were cut in the watertight bulkheads. A grand staircase required a spacious opening at every level, making a watertight deck impossible.... Stokers could work more efficiently if longitudinal bulkheads were omitted... A double hull ate up valuable passenger & cargo space; a double bottom would be enough."

"One by one the safety precautions that marked the Great Eastern were chipped away in the interests of a more competitive ship.... When the "unsinkable" Titanic was completed in 1912, she matched the Great Eastern in only one respect: she, too, had 15 transverse bulkheads."

"But even this was misleading. The Great Eastern's bulkheads were carried 30 feet above the waterline; the Titanic's, only 10 feet."

"The Night Lives On".

The Titanic's designers thought her quite safe enough, because she could float with any two of her 16 compartments flooded, and only the worst possible accident, a collision right at a bulkhead, could even flood two. Indeed, at the bow where the ship was narrowest and the compartments much smaller, she could float with the first four flooded, and collisions were most likely at the bow.

As the Titanic nudged and shouldered her way past the huge iceberg, we can now estimate that a gash was torn in her almost 100 metres long. It was probably more an irregular series of holes and rips, but the cumulative area along that great length was square metres. The ship's builder, Thomas Andrews, was aboard and inspected the damage with the Captain. They found that the first five compartments were flooding rapidly, and the sixth leaking.

Andrews quickly visualized the awful, inevitable mathematics. As the front compartments filled, and the bow sank, the transverse hull between the fifth and sixth compartments would drop over 10 feet-below the waterline. The water would spill over into the next compartment. So the ship would sink further, and water would spill into the next, and the next-and the pumps could only slightly delay it. The accident that nobody could imagine had occurred just that simply. Andrews underestimated her remaining time at only an hour. She lasted two.

By contrast, 50 years earlier on August 27, 1862, the Great Eastern had scraped on an uncharted rock off the coast of Long Island. It ripped a gash in her skin some 9 feet wide and 83 long, worse in some ways than the breach in the Titanic. However, the Great Eastern's inner hull was unbroken and the engine room remained dry. She not only floated, but limped into New York the next day under her own steam. Not a soul was hurt.

Engineers today, who work in such safety-conscious designs as nuclear plants, use the military term "defence in depth". Behind the first safety system lies another, and behind that, still another...each with its own backups. The Great Eastern had defence in depth against hull breach. By the era of the Titanic, liners had contented themselves with but a single "layer", the all-too-short transverse bulkheads. Soon after the disaster, the sister ship Olympic, and many other liners with comparable designs, were being expensively retrofitted with an inner, second hull. Suddenly the "impossible costs" of such "extravagances" seemed affordable after all.

Various other corners ceased to be cut in safety standards, as well. Since all lines did the same, competitive positions remained relatively the same. (Of course, the White Star line never recovered from the loss of the Titanic and the settlements for cargo and loss of life; it was absorbed by Cunard lines some years later. So much for competitive advantage from trimming standards.)

The most dumbfounding cut in retrospect was the lack of lifeboats. It was not just the price of the lifeboats themselves that bothered the businessmen, it was the deck space they ate up...one of the most precious commodities aboard. The committee of the British Board of Trade that made the regulations on life-boats was dominated by shipbuilders. They proved very able to convince themselves that boats for every person were not necessary. Thus the regulations of the time required only that a ship of Titanic's size carry boats sufficient for 962, though she could potentially carry over 3500 passengers and crew.

White Star's General Manager Harold Sanderson pointed out that the North Atlantic was so stormy that boats could not be lowered safely 95% of the time, and even once down, the passengers would be subject to additional dangers on the tossing sea. "They could avoid all this by drowning at once" joked the magazine "Fairplay", when he continued with this view even after the accident.

Needless to say, following the disaster, complicated formulas requiring so many cubic feet of lifeboat space per thousand tons of ship were replaced with a simpler one: enough seats for everyone aboard. Again, a supposedly high cost was suddenly affordable, and has never been questioned since.

The lifeboat problem was exacerbated by poor procedure. Only at the last did lifeboats leave full; at first, many left partly empty because passengers were not queued up to them. Second Officer Herbert Lightoller took the instruction "Women and

Children First" so literally that he let lifeboats leave with empty spaces rather than let men or boys as young as thirteen aboard...and was never so much as reprimanded for this part in seeing just over 700 saved when 1200 could have been.

Those of us who design and operate public services, and bear the title of "civil servant", may be much sobered by this transcript from the inquiry. Captain Maurice Henry Clarke, the inspector who cleared the Titanic for sailing, was being examined on the reasons for Titanic's only "lifeboat drill" having been conducted at the dock, consisting of only two boats, manned by hand-picked crew. Having conceded that he had since tightened requirements, he was asked:

"Did you think your system was satisfactory before the Titanic disaster?"

"No, sir."

"Then why did you do it?"

"Because it was the custom."

"Do you follow a custom because it is bad?"

"Well, I am a civil servant sir, and custom guides us a good bit."

Custom still does, and not just for civil servants. A few decades ago, seat belts in cars were thought a little-needed luxury; today vastly more expensive airbags are becoming a standard feature. Those who make and sell vehicles fought the transition step-by-step as the added cost naturally reduces the number that can be afforded by the buying public. The public itself has either resigned itself to the cost, or embraced it with enthusiasm, once the statistics of death and crippling were considered. The first hockey players to use helmets were jeered by other athletes until it became the new "custom". Now most parents willingly accept the cost of helmets simply for their children to ride bicycles.

All risks need rational consideration, and some must be accepted. Sooner or later, it will be pointed out that a few head injuries occur to pedestrians, but I hope that it will never be made mandatory to wear a helmet to take a walk. Even today however, it is still often the case that money management wins out over risk management. What this lead to with the Titanic was that a lesson was only learned because the tuition was paid in blood. Has much changed?

This week's news was full of the story that tractor-trailers are now being carefully examined for defects in their wheels and many pulled off the road, which costs money. This, after two people were killed by runaway wheels broken loose, both in the same week. There had been other such cases...but then two, close together, stimulated realization of a need for systemic change, not just a focus on the specific trucks involved.

I've been typing for hours, and my watch says it is 2:10. This is another time I recall from my Titanic reading. The ship went down about 2:20, and 2:10 was when her builder, Thomas Andrews, was last seen. After helping to organize the confused lifeboat loading, and personally assisting many to the last boats, he remained aboard. A survivor recalled seeing him at this time, alone in the Gentlemen's First Class Smoking Lounge, staring into space. The steward called to him to come try to swim for it, but he did not answer.

This room was one of the most remarkable on the ship, paneled in rich mahogany inlaid with mother-of-pearl, lit through stained glass, and even possessing a working fireplace. (No expense was spared here!) Andrews had done much of the architecture and interior design, as well as the structural and mechanical work.

I sometimes like to think that in his reverie, he realized the basic mistakes that he and a half-century of his predecessors had made. I hope it comforted him that this disaster was going to be so traumatic as to bring about tremendous changes despite all their costs, changes that would save more lives in the long run. Were it not for the Titanic, the safety standards might have continued for a long time, causing a long string of smaller disasters-each too small in its own right to bring about basic changes to the whole industry.

As the sinking bow lifted the stern a hundred feet in the air, Titanic's own weight broke her back and ripped her in two. The enormous screaming of tons of metal tearing was a sound that haunted the survivors all their lives. I hope Andrew's thoughts had given him some sense of consolation when the hungry North Atlantic stalked into the lavish Georgian drawing room to take away his sorrow and his shame.

Disaster Relief — from Ships

The U.S. Maritime Administration has struck a deal with the Transportation Command to provide ships for "humanitarian emergencies"-as set out in this MARAD press release.

The U.S. Department of Transportation's Maritime Administration announced today that it has signed an agreement with the U.S. Department of Defence's United States Transportation Command to use National Defence Reserve Fleet (RRF) ships, particularly DOT's Ready Reserve Force, to support government response to civil emergencies at home, as well as for defence sealift and humanitarian response overseas.

The agreement, signed by Maritime Administrator Sean T. Connaughton and General Norton A. Schwartz, commander of U.S. Transportation Command, formalizes using the ships in ways that have been tested under emergency conditions. During the hurricane season of 2005, RRF ships in Beaumont, Texas, provided the template for one use of the ships: emergency vehicles were driven on board two RRF ships berthed in Beaumont, where they were sheltered from Hurricane Rita, and were thus ready to respond to local emergencies when the storm was over. Other RRF ships provided food and shelter for recovery workers and first responders on the Gulf Coast during Hurricane Katrina and its aftermath.

Since its inception in 1976, the primary purpose of the RRF is to provide the U.S. Transportation Command with strategic sealift support for combatant commanders. The 44 ships of the RRF are owned by the Maritime Administration, crewed by civilian merchant mariners, and managed by commercial U.S. companies. The new agreement details the procedures the agencies will use for civil contingencies and national disasters such as the safe storage of emergency vehicles, providing fuel for local emergency services, or berthing and feeding of first responders.

This idea is not a new one. It has previously been discussed here, here. Also worth visiting is Professor Whitehurst's site here, which features his papers on "Hurricane Relief from the Sea."

Surveying Crewmembers of a Shrinking Ship

With a business community joined at the hip a hiccup in one market can influence the world economy.

The world is getting smaller. We have all heard it. The globalization of commerce, manufacturing and markets has led to the ever-increasing "shrinking" of business relationships and world visibility in the past 20 years.

The result of the togetherness is a business community that is operationally and financially joined at the hip with one another so much so that a hiccup in one market can result in a global butterfly effect that can influence the world economy.

Imagine then what would happen if a major, catastrophic loss hit one of these global markets, what impact it would have on the economies and businesses around the world. Not so hard to just imagine any longer.

When Ernst & Young and Risk & Insurance® Magazine set out to develop the second annual Catastrophic Claims Survey in 2011, it was without the knowledge one of the world's largest economies would suffer a crippling blow from a massive earthquake and tsunami.

Nearly 250 risk and insurance professionals from various industries were queried on their experience with claims, disaster recovery and — most notably — supply chain risk.

It wasn't until after the response to the survey that the events in Japan cast a bright spotlight on the true and real risk that a global economy and a global supply chain can have on businesses and the insurance industry overall.

There are many facets to a supply chain that cause companies around the globe to assign teams of people to identify, optimize and mitigate supply chain risk.

Whether it is the study of logistical patterns of shipping operations to squeeze pennies off each delivery or the identification of backup resources should one supplier fail to meet next year's supply contract, companies spend thousands of hours a year studying and analyzing supply chain risks. In fact, more than 70 percent of respondents indicated their company's supply chain is critical or very critical to its operations. The importance of a supply chain goes beyond those personnel that have to deal with suppliers, vendors, logistics and customers on a daily basis. Survey results show that senior management is very concerned with the issue—rightly so and deal with it on a regular basis. In Chart 1, more than half of the respondents indicated their senior managers address supply chain risks at least monthly.

For those respondents that deemed supply chain as "very critical," the percentage that address it at least monthly jumps to more than 80 percent.

Supply Chain Risk and Insurance

Risk management's role in evaluating and addressing supply chain risk varies widely. For some companies, risk management is critical to the success of communicating with the business to understand where contingencies and interdependencies pose risks to the company.

However, more often it is left to operations or even individual business units to understand the risks present and manage around them via operational adjustments to the business.

In fact, less than 25 percent of those surveyed stated that risk management played a key role in identifying and mitigating supply chain risk. This result is eye-opening considering that regardless of the plans that were made, adjusted and revised there are some risks that are simply unpredictable or unavoidable.

In response to this, the insurance industry has recognized the risks associated with their policyholders' complex supply chains and global interdependencies.

To help mitigate this risk, many property policies include contingent time element or contingent business interruption endorsements that specify what areas of a company's supply chain it will cover i.e. vendors, suppliers, customers. The broadest of these endorsements even allow for the effects of catastrophic loss and damage to second-tier or indirect suppliers and customers.

Considering the long history of property insurance, these endorsements are relatively new and only about one-half of companies have this tool in their risk management arsenal.

Given the recent global events this will likely change in the months and years to come.

The Japan Effect

So many superlatives have been used to describe the events in Japan, and they don't begin to adequately describe the effect these events have had on the country and its people.

From a business perspective, the rippling effects of the third-largest economy in the world taking a direct hit are still being felt globally. Companies all over the world are facing either their own loss of physical assets or the indirect effects of the lingering repercussions of these events.

Ernst & Young and Risk & Insurance® surveyed an additional 375 risk managers to determine how companies were affected by—and are addressing the crisis in Japan. Of those surveyed, nearly one in every three respondents indicated that they either suffered direct loss or had suppliers or customers that sustained damage.

It is no surprise that those industries hardest hit include technology and automotive companies, but others including consumer products, financial and professional services also are realizing the impact to their customers, suppliers and their own people.

Valuable resources, time, raw materials, and information have become scarce as Japan and its people focus internally to recover from such an epic disaster.

As the country begins the process of assessing damage and rebuilding its infrastructure and economy, so too are companies worldwide evaluating how they will find alternative sources of supply and demand for their products.

A complex interplay between suppliers, customers and companies' own interdependencies has often yielded more questions than answers, as to how businesses have been damaged or how they will recover. Of those surveyed, more than 60 percent of respondents indicated they were either taking a "wait and see" approach or are still gathering data and information to assess the impact to their business.

Many Japanese companies and operations were so severely damaged by these events they will never recover. This will leave gaping holes in supply chains and business networks that will have to be filled over time.

How companies will ultimately fare and recover remains to be seen. It is neither an easy fix nor one that will happen overnight. More than half the respondents that sustained some sort of loss from these events replied that they will continue to feel the effects for more than six months, and one in five for a year or longer.

Contingent Time Element

Those that have ever dealt with a business interruption claim know the complexities that come along with it. Assessing how the financial results were affected, the expected results had no loss occurred, how results at other locations may have been affected by the events, and tying all of that back to physical damage as the driving force of loss. Take all of that and assume your company sustained no physical harm and you have a recipe for confusion on how to approach this type of claim.

Contingent time element claims require information from and communication with suppliers and/or customers and often with indirect suppliers and/or customers. Understanding how any of the above was affected by a catastrophic event is critical to the proof and ultimate success of such a claim, as it is incumbent on the policyholder to reasonably demonstrate that its lost income—or increased costs—are a direct result of loss insured by the policy. Gathering adequate information about damage to indirect suppliers and customers can be extremely challenging, since there is usually not a standing relationship or open line of communication between the insured and its tier-two suppliers.

Often businesses do not even recognize they have a covered contingent loss because it is never brought to light internally. A supplier goes down, an alternative supplier is identified, albeit at a higher cost, but personnel directly involved think it is just part of running a business without the knowledge that the company may have insurance coverage for this type of event.

Ultimately, this may cut into the profit margin derived from the product or business unit, but it is either absorbed in the noise of rising costs, or worse compensated for by cutting costs elsewhere. Given this, the results of Chart 3 may not be surprising.

The events in Japan and the resulting claims will shed light on how policyholders and insurers will interpret and attempt to apply contingent time element coverage to their businesses. The outcome of these claims over the next several years will further shape the future of contingent coverage and resulting claims.

Supply chain risk is not getting any easier. However, with the proper level of communication, risk management can be the key to understanding the dependencies that business units, products or services carry as part of their ongoing operations. Often it is through this communication and education of the operations about potential risk and the opportunity for recovery that risk management will identify potential problems and solutions for mitigating against previously unforeseen disasters.

Risk Management: Unique Challenges

All commercial ships face the same risks, but to varying degrees, according to Commander Wilford R. Reams at the U.S. Coast Guard Cruise Ship National Centre of Expertise in Fort Lauderdale. These include fire, flooding, piracy, security and the environment.

In addition, cruise ships are subject to some unique risks, such as crowd control and crisis management issues, based on the large number of people onboard, Reams said. However, STCW (standards for training certification and watch-keeping) provides a basis for safety and people management, he added, noting that cruise ships have staff dedicated to crisis management, including a safety officer.

The magnitude of risk factors goes up as ships get larger, noted Reams, with more equipment onboard, more passengers and more crew. In addition, the cruise industry is pushing the envelope on innovation, coming up with the newest and greatest designs.

"Innovation can outpace prescriptive safety standards," Reams said. "There has been a shift at the IMO to goal-based standards and performance-based designs. When designs deviate on a high technical level, that means more extensive fire and evacuation modeling."

Other risk factors cruise lines must cope with are passenger safety and comforts on a more basic level, including slips, trips and falls.

As for fire, ships are built with active and passive systems, Reams explained, with the passive system being preferred, meaning structural boundaries that will limit a fire within a defined space for a certain time. Active systems are sprinklers that rely on sprinkler heads and pumps to work.

"You cannot design out all risk in any industry," Reams said. "However, cruise ships are subject to stringent and thorough layers of safety, including international instruments (SOLAS, STCW, etc.); flag state administration and class surveys; Coast Guard and U.S. Public Health examinations; as well as internal, company audits; and safety and risk management and training."

Cruise Ship Food Safety Inspections and Outbreak Prevention

We serve our clients with a full array of services designed to cost effectively meet the requirements under the Centres for Disease Control & Prevention (CDC) Vessel Sanitation Program (VSP). We offer comprehensive services in the areas, starting with construction planning, plan review, and shipyard inspections and continuing to communicable disease outbreak prevention and control, food and waterborne illness prevention, risk management and general environmental and public health.

Preventing problems and maintaining clean and safe facilities is a continuing challenge. Our diverse services are designed to prevent problems before they occur and to promptly correct any that may arise, as well as successfully prepare for CDC inspections. We guide our clients toward clean, safe, and sanitary facilities.

Construction and Plan Review Services

With our trained staff, which includes a former U.S. Public Health Service VSP Officer, we can guide you through the construction standards, plan review, and pre-inspection prior to the VSP CDC's shipyard inspection. This service by EHA can have the effect of a quicker and smoother PHS inspection with the net savings of significant time and money.

Inspection Preparation Services

We utilize rigorous inspection techniques, based on the CDC's VSP, to identify problems, educate employees, helping to avoid re-inspections, and solve sanitation problems before they lead to an outbreak scenario. EHA consulted with the CDC to develop the 1990 VSP and has a staff member with VSP program experience.

We incorporate HACCP principles focusing upon temperature control, cross-contamination and employee hygiene, while assessing equipment/utensil condition, water/sewage systems, garbage disposal, pest control and structural components (walls, floors, ceilings, lighting and ventilation). Self-inspection is a key facet to successful compliance and risk-reduction for food and/or waterborne diseases.

We can provide sampling of air, food, water and dairy products for laboratory analysis. Results accurately determine wholesomeness/adulteration, shelf life, compliance with regulations and how well the crew is handling the food and maintaining a sanitary environment.

Disease Outbreak Prevention and Control

Assistance in developing an Outbreak Prevention and Response Plan (OPRP), plans must indicate trigger points that initiate a ship's response to cases of illness. It must identify overall outbreak management strategies and include detailed infection control procedures for each area of the vessel.

After a potential link with a communicable disease outbreak, a skilled infectious disease epidemiological surveillance and investigation is essential. Our epidemiology and environmental health team includes former regulators, health department and VSP officials. We perform an independent review and trace-back for each individual case and cluster. We pursue the pertinent documentation and information.

Alternative causes and contributing factors are explored. In coordination with the CDC, Health Canada, or any other agencies involved, we also monitor their investigations and findings. Too often, even well-intended investigations by public authorities lead to misleading or inaccurate determinations with disastrous results for the cruise ship's public image.

Crew Training & Certification

Our certified instructors can provide your organization with the nationally recognized ServSafe manager's certification course in English and Spanish. Our instructors use slides, videotapes and demonstrations to augment lectures and classroom discussions. Hands-on task-oriented sanitation training can also be provided for your galley officers and crew. In addition, training is available for:

- General Environmental Health Maintenance (proper maintenance in child activity centres and ventilation systems);
- Potable Water (protection and handling of);
- Swimming Pool and Whirlpool Spas (maintenance of);
- Integrated Pest Management;
- Housekeeping and Infection Control;
- Waterborne Illnesses (both potable and recreational); and
- Cross-Connection Controls.

Crisis/Litigation Support

EHA has the experience and expertise to interact with the CDC, Health Canada, local and state health departments or other potentially-involved government agencies. In addition, we provide the required expertise for legal, insurance, and related proceedings. Members of our staff have been fully qualified in a variety of forums to testify as legal experts.

What You Need to Know about Maritime Security

Crime against ships is the largest concern for maritime security professionals and sailors operating in high risk areas.

Maritime security is as old as shipping, but modern pirates have advanced along with technology. The startling rise in major incidents is fueling a burgeoning maritime security industry. Do governments and the business community take this threat seriously? Do they understand the impacts of piracy at sea?

A Maritime Security Professional's View of Piracy

If the actions taken against ships and crews occurred on land we would refer to them as criminal acts. Maritime security issues are sometimes diminished by people outside the industry.

Recently a report was released that showed the economic impacts of breaches of maritime security. An estimated 12 billion dollar impact to the shipping industry annually was reported by the authors. This information, when reported by a reputable publication, was prefaced with the exclamation "Arrr!". This sort of mockery of crime is considered completely unacceptable when applied to other despicable acts, so why is it so well tolerated?

Our image of pirates is formed by books, films and Halloween costumes. Maritime security professionals see the true nature of a modern pirate; aggressive, violent, and desperate. The beloved singing, dancing, wisecracking pirate only exists on the movie screen and in theme restaurants.

What are the True Costs of Lapses in Maritime Security?

The answer to that question depends on who you are.

If you work on board a vessel operating in a risky area the costs might not always be economic. Psychological strains from perceived threats and increased diligence can cause fatigue and a reduction in efficiency. If your ship is overtaken by pirates the costs may be petty theft, extended detention, or loss of life.

Shipping companies faced with maritime security issues absorb much of the economic cost.

Training and equipment to repel attackers is costly. Each vessel must be outfitted with stronger physical security features and devices designed to deter unauthorized boarding. Every person on board must undergo security training.

The untrained crew member will certainly be taken hostage more easily, and once one hostage is taken the attackers have the advantage.

Ship operators also have added cost from increased fuel consumption because of additional sea hours due to longer routes to avoid troubled areas. Higher rates of speed used to minimize time in a danger area can also lead to more fuel costs.

Insurers must charge a premium to their customers who are operating in zones of reduced maritime security. Direct threats from pirates contribute to the increase in premiums but the riskier modes of operation used to avoid pirates can contribute to more traditional marine accidents.

As vessel security improves in the future it would not be a surprise to see pirates intentionally wreck a ship to extract a relatively small amount of loot for their risk. Salvage and environmental costs could make today's costs seem tiny.

Law enforcement and prosecution of seagoing criminals by governmental bodies is a large piece of the financial impact. Pirates depend on operating in a sparsely patrolled environment; the vast area which needs to be covered gives the criminal the advantage. A fully equipped and crewed military ship requires significant direct resources in addition to support services for the vessel. Some solutions, like remote escorts, have not yet matured technologically or politically. Now add the hours and funds needed to develop and deploy maritime security agreements and you see the total rise even more. This, of course, is just the starting point. Consider some of the impacts that are harder to quantify like loss of economic development, reduced market opportunities, and national animosity towards sailors and enforcement personnel and the negative impact becomes clear.

The greatest effect could be on the consumer. By all accounts the majority of consumer products or the raw materials they are produced from travel by ship at some time. Costs will certainly be passed to the end user.

If Maritime Security is Such a Problem What Can be Done?

Better Physical Security

Innovative solutions and affordable deployment of technologies is essential to the security of ships. Devices must be robust and immune to countermeasures by attackers.

- Optical Dazzlers are used to temporarily blind pirates before they can approach.
- Sonic Pain Devices are available to target approaching threats but are also used as a deck mounted obstacle once attackers have boarded.

- Supplemental Radar can detect suspicious craft which is following in the main radar "blind spot".
- Attack resistant deck rails; some operators attach razor tape to the deck rails and manually fend off boarding hooks and ladders. Safer alternatives must be developed.
- Locks; spaces must be secure but need to be quickly opened for escape in case of emergency. Devices must comply with existing regulations.

Effective Policy and Prosecution

- The authority of international maritime organizations must be increased.
- National policies specifically aimed to eliminate modern piracy must be implemented. Recently an outdated piracy law in the United States allowed a captured pirate to avoid prosecution.
- The policies concerning possession of firearms by private firms in international waters should be enforced or modified to address the current situation.
- Piracy in international waters must be prosecuted in international piracy tribunals.

Social Deterrents

- Efforts to eliminate the root causes of piracy in affected countries could yield the greatest reduction in incidents. Somali Pirates see piracy as an escape from poverty and social turmoil in their country.
- Cultural integration of pirates within society must be avoided. Pirates need to be portrayed as criminals, not benefactors of their communities.

How to Negotiate With Pirates

On September 26th 2008 a Ukrainian cargo ship was seized by pirates 200 miles off the coast of Somalia. The cargo aboard consisted of battle tanks, rocket propelled grenades, and anti-aircraft weapons. The ship was held for ransom. More than four months later pirates escaped with over three million dollars but no weapons.

Avoid Negotiation With Pirates

Even with advice from experts, the security plan of a ship may fail and allow pirates to board a vessel. The number and sophistication of pirate attacks each year continues to rise. Many of the attackers are deterred by well trained crews using existing shipboard equipment like fire hoses. Most of those that manage to get on board steal what they can from the ship and crew then depart quickly.

According to monthly Piracy Information Reports published by the IMO; hostage situations do occur but are a small portion of overall reported incidents.

What if you are Forced to Negotiate?

Some situations require negotiation. In the case of the Ukrainian vessel it was unacceptable to allow pirates to obtain the arms and ammunition on board. The military was faced with a choice of attacking the vessel and almost certainly causing casualties or allowing the demanded ransom to be paid.

Just one captured seafarer, when threatened with harm, can become the bargaining tool pirates use to take control of a ship.

The bulk freighter "Bet Fighter" a U.K. flagged vessel was attacked in August 2010 off the coast of Palau in the South China Sea. Six pirates armed with "long knives" were able to take control of the vessel by gaining access to the bridge. The Master was captured by pirates in his cabin. The pirates fled after stealing cash and property from the ship and crew.

Even though the pirates who attacked this ship had control of the bridge and Master they fled with little reward for their efforts. The need to negotiate depends on the goals of the pirates.

Know Your Enemy-Like all criminals, pirates have different definitions of success. Some are satisfied escaping with paint, lines, or safety gear; others set out to steal cash. Few seek a protracted ransom situation that may last for months.

Using reported information a Master can determine what is most commonly being stolen in a high risk area. If pirates do get on board the ship and take control it is best to know what they expect to gain.

Knowing the shore situation is also important. If a local resident earns the equivalent of $350 U.S. per year which is the case in Somalia. It is reasonable to think pirates will be satisfied with a small amount of cash and supplies. If they find valuable cargo, like the arms on the M/V Faina, demands will certainly increase.

The most dangerous situations occur when the pirates aren't able to satisfy their desires. Negotiating for the Safety of Your Crew-Pirates who ransom ships use the safety of the crew as leverage. The informed Master and crew will be able to control their situation until a course of action can be determined and executed by responders. Negotiation is a form of communication that can be verbal, but non-verbal actions are also a form of negotiation. Be careful about what you say and do.

Strategies for the Master

- Make sure your crew knows they aren't expected to be heroes. An individual who attacks captors endangers everyone.
- Allow the pirates to feel in control. People under stress are more likely to act irrationally and harm your crew.
- Establish a regular routine for yourself and others. A routine is comforting, relives the stress of the unknown, and allows responders to plan a response more easily.
- Mentally record details of the pirates. Subtly note appearance, command structure, clothing and equipment, subjects of conversation, and other things that may prove useful in negotiations. The information will also be valuable to investigators.
- Be Patient.

Strategies for the Crew

- Don't antagonize your captors. Be cooperative. Sômeone is negotiating for you, don't undo their hard work.
- Remember details about the pirates to share with investigators later.
- Establish a routine. Eat, sleep, and groom yourself as regularly as possible. Try to include some distractions like

reading or exercise. Some ransom situations last for months; your routine will help time pass more quickly.

- Find areas of mutual interest with your captors. Respectfully relate your desire to be released.
- Be Patient
- When responders arrive get on the floor and stay there until the situation is neutralized.

When the Negotiations are Over-Be sure to get some rest and confront the experience as soon as possible. Being a hostage is traumatic, accept professional counseling if it is offered. The effects of stress may not appear until long after the incident is over.

Blackwater Security's Anti Piracy Plans

Anti-Piracy Plans Detailed

The diplomatic documents released recently by the WikiLeaks organization include a confidential cable from the US Embassy in Djibouti. The message was addressed to several intelligence and administrative units in the US government.

Although the subject line of the cable is "Djibouti Approves Blackwater for Commercial Counter-Piracy Operations" it is clear the embattled Blackwater Corporation considered an operational plan that was unprecedented in commercial operations.

The private company's plan included the use of an ex-NOAA ship named the "McArthur" to escort up to three commercial ships at a time through high risk areas. Armament aboard the 138 foot ship was to include .50 caliber machine guns.

The crew of 33 US citizens included a total of 18 security personnel. An unarmed aerial surveillance device was also mentioned by the embassy communication.

The International Maritime Organization advises against arming commercial ships with lethal weapons. Armed military escort ships from several nations patrol the east coast of Africa and have responded with lethal force in some cases.

Blackwater's strategic operations plan stated that it did not intend to take any pirates into custody, but would use lethal force if necessary.

A Djibouti government official who filled several security roles gave permission for the paramilitary operations to be based in the Port of Djibouti. Provisions for the security of the ships weapons while in port were to be provided by Djibouti Navy.

Embassy officials expressed concerns about the legality of the plan in the February 12, 2009 document. The diplomatic post also requested advice on their involvement with the planned high profile launch of the service in March of 2009.

After four years of investigations and criminal prosecutions by the US Government the company, which is now known as XE Services, has not been held accountable for violent crimes in Afghanistan and Iraq.

Will a Sale Put an End to Anti Piracy Efforts?

The sale of the organization was expected to be announced in mid-December 2010. A group of Los Angeles based investors and a New York private equity firm reached an agreement for purchase of XE Services from the former president of the company, Erik Prince. The terms of the sale are confidential but it is known that a portion of the proceeds of the sale are based on future performance.

Mr. Prince is unlikely to have any influential role in the company although he retains some ties and continues to receive income based on XE's earnings.

According to the US State Department, no further contracts would be awarded to the firm while Mr. Prince was involved with operations. Civil lawsuits and federal weapons charges have been filed against executives of the company including Prince. Court records show that Erik Prince has now relocated his family to Abu Dhabi.

One of the primary investors in XE is Jason DeYonker, of Forté Capital Advisors. DeYonker advised Erik Prince while he was developing Blackwater's business plan and managed the finances of the Prince family for several years. The sale price is expected to be around 200 million dollars.

The new owners plan to shift the focus of the company from private security operations to training for military and police forces. Fines imposed by the US Government for export violations will continued to be paid under terms of a settlement. Payment of the 42 million dollars in fines will allow the company to receive future government contracts.

Anti-piracy operations reported in embassy cables released by Wikileaks are unlikely to move forward considering the company's reluctance to engage in private security operations in the future. It may be possible for the company to profit from anti-piracy operations by training national forces of affected countries in defensive techniques. The controversy of armed private firms patrolling areas of pirate activity would be eliminated by enlisting and training forces tied to nearby nations.

An increase in pirate attacks since Blackwater abandoned plans for their private security patrol in 2009 shows that a solution to this problem hasn't been found. Pirates continue to attack large and small vessels in order to take hostages and gain ransom payments. Considering the resources being expended, it is simple to conclude that a firm which can provide a solution will be paid handsomely.

Currently there are few options for the resolution of hostage situations other than the United Nations recommendation of effective negotiations.

Cruise Vessel Safety and Security Act of 2010

The Cruise Vessel Safety and Security Act

This act establishes regulations for the recording and reporting of crime and accidents aboard cruise vessels in waters not previously subject to United States law. The law also requires new physical security devices on existing vessels and increases requirements for new vessels. All parties should refer to the full text of the law (.txt),(.pdf) and consult their legal professional to determine proper implementation.

What Vessels are Subject to this Law?

All vessels authorized to carry at least 250 passengers, and provide sleeping accommodations for each passenger. Also, any vessel which embarks or disembarks passengers in the United States is subject to the Cruise Vessel Safety and Security Act.

Retrofit Responsibilities

To be in compliance with the new regulations a vessel must;

- Have "ship rails that are located not less than 42 inches above the cabin deck".
- Equip passenger and crew cabins "with entry doors that include peep holes or other means of visual identification".
- Integrate video surveillance systems to document crime or man overboard situations.
- Have the ability to broadcast audio alerts throughout the vessel.

New Vessel Design Requirements

Additionally, new vessels constructed after July 27, 2010 must;

- Equip each passenger and crew cabin entry door with "security latches".
- Equip each passenger and crew cabin with "time-sensitive key technology".

Who is Responsible for Compliance?

Responsible parties include; owner, charterer, managing operator, Master, or other individual in charge of a vessel.

On Board

Record Keeping

A log book or electronic record is to be kept in a central location that is accessible to law enforcement personnel. All major or minor crimes reported by passengers or crew are to be recorded. This record must be provided to any law enforcement officer who requests access in an official capacity within the scope of an investigation.

The information recorded must include;

- The vessel operator
- The name of the cruise line
- The flag under which the vessel was operating when the reported incident occurred
- The age and gender of the victim and accused assailant

- The nature of the alleged crime or complaint
- The time and date the incident occurred, if known
- The vessel position at the time of the incident or complaint
- The time, date, and method of the initial report
- The law enforcement authority to which the initial report was made
- The total number of passengers and crew members aboard the voyage
- Any case number issued by law enforcement.

Reporting

The responsible party will;

- Telephone the Federal Bureau of Investigation as soon as possible after the report of a serious crime or missing person
- File a written report of serious crimes to the Secretary of the United States Coast Guard via an internet portal provided by the Secretary. Lesser crimes and incidents may also be reported through this portal.
- Report any other criminal incident involving passengers or crew to proper local law enforcement authorities

Privacy

Each passenger reporting a serious crime on board a vessel will be given free and immediate access to a private internet terminal and telephone line to contact law enforcement, private attorney, or other third party support. All medical information is to be kept private, with exception of the Master of the vessel who is required to report this information to investigators, or to assure public safety on board the vessel. Any information regarding victim communication is also subject to privacy rules.

Passenger Information

Each passenger shall be provided a "security guide" written in plain language. The guide must include;

- A description of medical and security personnel on board, and 24 hour contact information for these resources.

- Law enforcement contact information and procedures for incidents occurring; in United States territorial waters, on the high seas, and in any country visited on the cruise.
- The location of, and contact information for, the United States Embassy and United States Consulate in each country to be visited in the course of the voyage. This information must be posted in each passenger cabin, and additionally must be posted in locations accessible to the crew of the vessel.

Crew Access to Passenger Cabins

Access to passenger cabins by crew members will be controlled by the responsible party. Crew members may not enter passenger cabins unless authorized according to job description or special permission. Permission logs and policy compliance logs should be kept and audited occasionally.

On Shore

Online

The passenger "security guide" is to be provided on the cruise line homepage.

Filing

A copy of the passenger "security guide" is to be filed with the Federal Bureau of Investigation.

Medical and Security Personnel

Supplies

Maintain in-date supplies of drugs to combat sexually transmitted disease in cases of sexual assault. Maintain the equipment and supplies to perform medical examinations in cases of sexual assault and to assure the preservation of evidence in such cases.

Training/Certification

The responsible party must provide, at all times, passenger and crew access to medical and security personnel that are trained and certified in the procedures needed to treat and document cases of sexual assault and other crimes. Responders will not disclose medical information, except to the Master for purposes of required reporting or passenger and crew safety.

When Will These Regulations Take Effect?

- This act was signed into law on July 27, 2010
- New construction rules were implemented on July 27, 2010
- Crew training certification will begin one year after implementation, minimum trained personnel must be on board as required two years after implementation.
- Safety equipment upgrades must be completed 18 months after implementation.

Are there Penalties?

Violations of these regulations may result in –

- Criminal penalties up to $250,000 in fines, one year in prison, or both
- Civil forfeiture up to $50,000 per violation.

What is the Cruise Vessel Security and Safety Act of 2010?

Answer

The U. S. Congress passed the Cruise Vessel Security and Safety Act of 2010 in order to improve cruise passenger safety and ensure that, should a sexual assault occur onboard a cruise ship, the victim has access to trained first responders and to a confidential means of communication with law enforcement, legal and victim advocacy professionals. This law applies to cruise ships with overnight sleeping facilities that carry 250 or more passengers that depart from or call at U. S. ports.

Enhanced Vessel Safety Features

Vessel owners must install the following safety features within 18 months of the law's enactment date:

- Peep holes in passenger stateroom and crew cabin doors;
- Ship rails that are at least 42 inches high;
- Video surveillance equipment for monitoring passenger safety and "passenger overboard" situations;

- Acoustic hailing or warning devices that can be heard throughout the ship, for alerting passengers to danger;
- Passenger security guides, which will explain the process for reporting a serious crime, with detailed information about the legal and law enforcement policies and points of contact applicable if the crime occurs in port (U. S. or foreign) or at sea. The security guide will also include material about medical and security crewmembers' responsibilities and provide contact information for these crewmembers.

Ships built after the law goes into effect must have security latches and time-sensitive key technology installed on each passenger stateroom and crew cabin door.

Cruise lines must set up procedures for determining which crewmembers may enter passenger staterooms and documenting the permitted access times.

Trained Medical Personnel

Every ship's staff must include a doctor or registered nurse who is trained in assisting sexual assault victims and preserving evidence for use by law enforcement officers. Ships must carry anti-retroviral medications. Trained medical personnel who know how to administer these medicines must be onboard each ship, and a certification program will be developed and implemented to ensure that this training is adequate. Victims must also be able to contact legal, law enforcement, embassy, Coast Guard and victim support service offices via private telephone or Internet connection from onboard the ship.

Cruise Vessel Crime Reporting

Each ship must create and maintain a log of crime reports and track serious crimes (murder, suspicious death, missing person, kidnapping, assault with serious bodily injury, theft of items valued over $10,000, tampering with the ship or setting it on fire, sexual abuse or sexual abuse of a minor), thefts of items or money valued at over $1,000 and all other reported crimes. The statistics must be reported to the U. S. Department of Homeland Security for inclusion in a database that is accessible to the public.

If a serious crime takes place on the vessel, it must be reported as soon as possible to a field office of the Federal Bureau of Investigation (FBI) or to a legal attaché and reported in writing to the Department of Homeland Security.

Effective Dates

Most of the provisions of this law go into effect one year after the date it takes effect. The certification of medical personnel training takes effect two years after enactment of this law.

Maritime Law by Barb Karg and Arjean Spaite

Maritime law is defined as the body of law relating to maritime commerce and navigation, and to maritime matters in general. Each country generally controls the waters surrounding it, and their laws apply, but there are vast areas on the high seas that aren't owned or governed by any country. Because of this, the international community has often attempted to work together to form laws that would cover the world's oceans. In 1958, the United Nations Conventions on the High Seas was held, and in 1982 the Convention on the Law of the Sea was convened. These two groups defined piracy as "an attack mounted for private ends on a ship on the high seas that involves violence, illegal detentions of persons or property, or the theft or destruction of goods." The more recent United Nations Law of the Seas defines piracy as "any illegal acts of violence or detention, or any act of depredation, committed by individuals for private ends against a private ship or aircraft." They also specify that two vessels must be involved in the incident, otherwise the act is one of hijacking rather than piracy.

Why was there a need for a Change in the Definition of Piracy?

There are problems with the first definition, including the fact that it states that the attack must occur "on the high seas." Many pirate attacks are much closer to shore, some even occurring on docked boats and rivers. Also, this definition said that the attack had to occur for private gain, and this is difficult to prove given that pirate's identities are typically unknown and the individuals can't be shown to be acting for their own private gain.

The majority of modern pirates operate out of a specific homeland, and piracy is most common near areas where authorities tend to turn a blind eye to their acts.

Most countries have differing levels of laws and law enforcement, and local governments may condone or even participate in acts of piracy. For example, in 1995 the Philippine Coast Guard boarded the merchant ship Kafu Maru, allegedly searching it for illegal firearms. They found no firearms, but did commandeer the ship's cargo of cigarettes and liquor, which was worth over 4,500,000 pesos. The coast guard officers neglected to report that they'd taken the cargo, despite the fact that it was later found in their homes.

Outside of local waters, international maritime laws prevail in any instance of piracy. If a ship on the high seas is suspected of being a pirate ship, it can legally be boarded and seized by any official government vessel. The pirates can then be arrested, and subjected to the laws and penalties of any country that apprehends them.

Cruise Ships: A Playground for Fun or Playing Field for Malevolence?

Where we have Been

Not Safe at Sea

During 2009, 25 people went overboard on cruise ships, many endured sexual assaults and robberies, while others have just "disappeared." Cruise lines have taken the position that they have no legal obligation to investigate crimes and do not have the technical staff to examine the "alleged" offenses. Since they have not been legally bound to report crimes onboard their vessels, they have "voluntarily" notified the FBI when Americans became victims. With the bureau miles from the incident, it is impossible for the agency or other law enforcement departments to take immediate steps to protect the victim, secure the crime scene, or start an investigation. Very often jurisdictional issues prevent the investigative process from even beginning.

The Government Accountability Office (GAO) cruise ship security report reviewing a Coast Guard study, recently determined that cruise ships are a target for terrorist boarding's and waterside attacks.

Almost Twice a Month

Assault statistics for Carnival Cruise Lines claimed that from 1993 through 1999, 108 incidents against both crew members and passengers occurred, including rape, unwanted touching, and unwanted kissing. The accused crew included a cabin steward, an assistant cook, supervisors, engineers, and even a chief security officer. As a result of the assaults, Carnival fired 47 employees over the five-year span, the standard "punishment" in such cases.

CLIA Represents the Industry, Not the Passenger

Between 2008 and 2009, the cruise line industry spent US$11 million for professional lobbying efforts in Washington to avoid legislation. The Cruise Line Industry Association (CLIA) represents the industry and has pushed hard to avoid legislation. In 2007, CLIA produced a 32-page report that offered no commitment to change. The report asserted that the crime rate for the industry was only .01 percent although the statement was untrue and the falsehood was confirmed in an April 4, 2008 letter from the FBI. Even CLIA's own expert witness at the Senate hearing, Dr. James Alan Fox, indicated that this was not a valid annual crime rate. It is really impossible to actually determine crime rate at sea as CLIA "officially" does not keep track of crimes.

Accountability: Kerry Bill

Thanks to the efforts of the International Cruise Line Victims Association (Internationalcruisevictims.org) directed by Kendall Carver, the Kerry Cruise Vessel Security Act of 2009 will increase security, law enforcement, and accountability on cruise ships operating in international waters. According to Michael Ehline, an LA-based personal injury attorney and expert on injuries at sea, the new legislation, "...will require log book entry and reporting of deaths, missing individuals, and alleged crimes as well as many other things designed to make a cruise safer."

In addition, the act considers passenger security as it relates to vessel design, equipment, construction, and retrofitting; the use of video surveillance to monitor crime; posting of US embassy locations; maintenance of equipment and medical personnel for sexual assaults on board; confidentiality of sexual assault information; and crew access to passenger staterooms.

The act also prescribes penalties for violations. The secretary of the department in which the Coast Guard operates is required to develop training standards for: (a) certification of passenger vessel security personnel; (b) crew members and law enforcement officials on methods for prevention, detection, evidence preservation; and (c) reporting of criminal activities in the international maritime environment. Vessels carrying US citizens are not permitted unless there is at least one crew member on board who has met training and certification. Persons who violate the requirements are subject to penalties. The secretary is also directed to study and report to Congress on passenger security needs depending on the numbers of passengers-with recommendations for improved security.

Baby Steps

The act takes a baby step toward making the cruise experience safer; however, there are many missing parts. To make the legislation more comprehensive, Attorney Ehline suggests that cruise lines doing business in the USA: (1) review their personnel hiring practices; (2) not hire people from countries hostile to American interests; and/or (3) do a complete and thorough background check on each and every employee – regardless of citizenship. Ehline noted that some countries offering employees jobs on cruise ships do not meet US minimum pay requirements and do not have the "ability to offer a comprehensive data supply to the cruise corporations who employ them."

The Coast Guard has implemented a few new procedures including: escorts of high-capacity ships (resources permitting), and the review of passenger and crew data.

Passengers: Profits Before People

While the new legislation may be helpful to American travellers, other countries have not been as responsive to the wants/needs of their citizens. According to Ehline, "...Americans have a tort system designed to make them whole when injured. Other countries, other than those operating under the English

Common Law Systems, simply allow a more harsh Roman Civil Law approach" which does not offer the framework with the same protections. "Americans take a lot of cruises," according to Ehline and a large market source are elderly passengers from Florida and California; unfortunately "they get injured... due to design flaws that don't take them into consideration... More people mean more injuries."

Weakest Link

Current cruise line hiring practices, according to Ehline, leads to weak links in the chain of safety. "I think the maritime framework with its treaties, the unfair venue provision in the passage contract (forcing an injured victim who lives in Georgia, for example, to sue in Los Angeles, California), the unfair claims filing ... requiring a wounded passengers to bring an informal claim within 6 months or you cannot sue...All these sidesteps and loopholes are just perfectly designed for those who would rather not deal in the system of laws..." commented Ehline.

Criminal and/or Civil Law

The cruise ship is a common carrier and Ehline determined that, "...there is at least one case that holds a cruise ship strictly liable to injuries caused due to the cruse ship's negligence or wilful neglect. In the case of a cruise ship employee raping a passenger or co-worker, issues at sea become a criminal issue. If the passage contract says you have to file a civil lawsuit in California State court, the same substantive tort rules for any another California citizen would apply." The victim is "...usually forced to hire a California lawyer, sue in California...and there are fees to fly in the doctors and other witnesses to and from California (plus the cost of hotels, etc.)". Ehline finds that cruise ships practice, "Draconian procedural rules." These rules often "...force an unfair settlement due to economic inability to sustain the lawsuit....and...is exactly what the cruise lines want," according to Ehline.

So – Sue Me

With Americans being a litigious group, the immediate response to any injury is to sue. According to Ehline, "Often, a good cruise ship injury attorney will be able to resolve claims without suing. But that is rare. My experience has been that suing is the best way to get their attention...The real victory is just protecting the claim itself.

Many a legal malpractice claim has arisen due to inexperienced attorneys assuming you have one to two years to sue a cruise ship (and not even knowing about the passage contract that abrogates those rules)."

Anticipation

Ehline offers suggestions for staying safe onboard.

-The safest decision? Don't go!

However, if you are bound and determined to take a holiday at sea:

1. Don't fraternize with the crew
2. Travel in twos at all times
3. Never leave your child unattended OR allow them to participate in any cruise ship sponsored activity unless you are THERE the entire time
4. Don't go on a shore excursion unless you read the passage contract first (cruise ships have some passage contracts that preclude them from liability for cruise ship sponsored activities of the ship).

If an incident occurs:

1. Immediately report in writing the incident to the security department on the ship
2. Get a copy of your written report
3. Take photographs of the scene of the incident before it changes, if possible. If the scene has changed, still take photographs
4. Write down the name, address, and telephone number of each person who is a witness to the incident
5. Go to the ship's doctor to get treatment for your injury
6. If necessary, go to a shore-side doctor at the next port of call and contact a Personal Injury attorney immediately.

Trends

With the notion that "my mind is made up, don't confuse me with the facts," more and more people are taking holidays at sea. Research continues to show that cruise ships out-score all other vacations for perceived value. Many cruise lines are seeing increases in groups including multi-generational, girls' getaways/man-cations, plus civic and social groups. Fredik Johansson, senior architect for Tillberg Design AB, (QE2, QM2, Crystal Serenity and various Disney ships) predicts "mother ships" that will launch fleets of smaller vessels and space-efficient ferries (PSA Annual Cruise Report, 2008). The Report also forecasts two million passengers by 2010

To further seduce travellers and their families, new onboard amenities are likely to include full-scale aqua-parks, luxury spas and spa suites, increased dining choices, areas dedicated to adults, teens or children, expanded golf programs, and enhanced Wi-Fi connectivity. Popular destinations include the Caribbean, The Bahamas, Alaska, Europe/The Mediterranean and Mexico.

Statistical Odds of Danger

Travel by sea can be as much fun as the advertisements lead us to believe. To keep safe, be prepared by assessing risks and have a viable plan to mitigate the possibility of something going wrong.

Cruise Ship Safety

Approximately 10 million Americans vacation on cruise ships each year. The majority of these people has a wonderful time and experiences all the perks that a cruise vacation has to offer such as stops in exotic locations, pampering, and top-notch food. However, a select group of passengers will experience the dark side of the cruise industry; an inebriated partier falls over the side and drowns, a female passenger is raped, or a norovirus outbreak causes illness amongst a boatload of travellers. Allowing yourself several minutes to view cruise ship safety tips can increase your chances of having a safe vacation. According to the International Council of Cruise Lines, from 2002 to 2005, 24 passengers turned up missing, 178 reports of sexual assault were filed, and four robberies occurred across 15 different cruise lines.

Victims of cruise ship injury and crime have recently called out Congress to heighten security regulations throughout the cruise industry. Since cruise ship medical staff and security personnel are employed by the cruise industry, the victims believe this creates a conflict of interest; they are unsure whether cruise line employees are out to protect the passengers' well-being or the financial interests and reputation of the cruise industry.

Experts still insist that cruises are one of the safest travel methods. From April to August 2007, just 207 (.01 percent of total passengers) incidents of concern were reported to the FBI, and a mere four of these were for missing people. Out of these 207 crimes, sexual assaults occurred most often, preceding assault with bodily harm and theft of property worth more than $10,000. Following certain guidelines can help enhance your cruise ship security.

When you first board a cruise ship, the cruise ship's captain will call everyone to the deck to do a mandatory review of emergency safety measures. It is important to pay attention during this procedure as you receive instructions about what to do and where to go in the event of an emergency. Although this information is extremely beneficial for passengers to comprehend, there are still some things that passengers should be aware of that the cruise line does not necessarily explain. These eight cruise ship safety tips will help keep you and your loved ones safe so that you can get the most out of your cruise vacation:

- Be logical. Just because you're on an exciting getaway doesn't mean you're not vulnerable. Cruises carry thousands of passengers, and although most of them are there to have innocent fun like you, not everyone on the ship has the best intentions. Avoid walking down dimly lit corridors, back down from confrontations, do not invite strangers into your room, and do not give out personal information to someone you don't know.
- Leave precious belongings at home. Various cruise ship workers (e.g., maids) can have access to your cabin. Valuables such as cash, plane tickets, passports, and jewelry may be stored in the cabin's safe, but if it has a bypass code, there is a chance that an employee may know it.

- Cruise ships are notorious for sexual assault crimes. Women should never leave a drink unattended or accept one from someone they never met before because it could be spiked. In general, all passengers should consume alcohol in moderation. Most altercations and falls are the result of overindulgence.
- If you go overboard, odds are you will not be found. Cruise ship personnel are required to conduct a considerable search for passengers who fall off the edge, but the chances that this person will be located are slim. People often go over the side when they are leaning or climbing on railings. You may be tempted to recreate Leonardo DiCaprio's famous scene from the movie "Titanic," but you are only putting yourself in unnecessary danger.
- The applicable laws governing cruise ship crime depend on the vessel's location. If a crime is committed while the ship is docked, the criminal is subject to the country's jurisdiction where the port lies. When the ship is out at sea, laws are based on the country who oversees these waters. However, if the ship is in a non-territorial section of the ocean, maritime law resumes.
- Serious crimes involving a US citizen are supposed to be reported to the FBI. It is at the discretion of the ship's security team what constitutes a serious crime. Crimes committed against an alien are not included in these reports. If you're the victim of a crime, find out if it is being reported to federal or local law enforcement. If the crime is report is withheld, you can take it upon yourself to notify the authorities.
- Over a period of a year-and-a-half, the cruise industry spent $2.9 million on federal lobbying. Many believe that the government has very little control over the present-day industry and may even fabricate their crime statistics. "I don't have a comfort level that statistics from the cruise industry are accurate. I've never known statistics that are voluntary to be accurate," US Rep. Christopher Shays told the Associated Press.

- Most cruise lines are based in other countries. When you see a ship waving something other than the American flag, it is partially governed by its own set of laws.

Although experts say cruises are still a very secure way to travel, the cruise industry has begun to acknowledge recent safety issues. For example, Royal Caribbean is drilling peepholes through all cabin doors, setting up suicide hotlines, and requiring all employees to take a course on sexual harassment. Following the above cruise ship safety precautions can help ensure that you and your family have a safe cruise.

Common cruise ship injuries include slip and falls, cruise ship doctor malpractice, weather-related accidents, fires caused by engineering accidents and food poisoning. The cruise ship may be liable for your injuries. If you have suffered a cruise ship injury, have been sexually assaulted on a cruise ship, or have experienced any other cruise ship accident, contact a cruise ship injury attorney immediately. Do not hesitate to reach out to an experienced cruise ship injury lawyer to receive the compensation you deserve.

Practical and Legal Tips for Safe Cruising

Summer is here, full of travel and memorable moments. To ensure smooth sailing this summer, consider the practical and the legal before embarking on a cruise.

Practical

First, it is important to note that just because you board a ship in a United States port, it does not mean you are protected under the United States legal system.

Often ships are registered in other countries and, once out at sea, U.S. laws may not apply.

In an effort to minimize crime and increase your traveling pleasure; here are some helpful tips, according to www.crimedoctor.com.

- Make sure your luggage has a lock on it, as this will deter theft.
- Inventory your luggage and create a list of the contents. You may even want to take photos as an added precaution.

- Use common sense when entering your cabin. Other crew members have keys to your cabin so it is important to check to make sure it is empty when you enter and use the safety box inside for valuables.
- Stay in a group or well-lighted areas while aboard your ship. Strongly monitor your teens and their whereabouts.

Safety and security are the goals for all Americans as cruise ship passengers. Presently, we are waiting for the President's signature on a bill which passed on June 11, 2010, the Cruise Vessel Security and Safety Act.

Federal Legislation

The Cruise Vessel Security and Safety Act was passed with bipartisan support with Congresswoman Doris Matsui representing the 5th congressional District in California and Sen. John F. Kerry from Massachusetts in the United States Senate.

The legislation would require the cruise industry to increase rail heights to 42 inches, increase cabin security with peep holes on all doors and have a sophisticated video surveillance system.

The video surveillance program would detail crimes committed aboard the ship, making prosecuting and documenting a crime much easier.

The bill would also mandate cruise personnel to contact the Federal Bureau of Investigation (FBI) and the United States Coast Guard in the event of a serious crime.

Serious crimes would include homicide, suspicious deaths, kidnapping, missing nationals and assault.

Further, rape kits and a sexual assault forensic expert would be on board each ship. The Secretary of Transportation, along with the FBI, will also establish a program to train all crew members in detection, prevention, reporting and preserving evidence on all cruise ships.

So, if you are planning a family cruise, request a copy in writing of how the cruise ship handles problems such as lost luggage through criminal acts. Remember, a safe vacation is a fun vacation!

4

Health Safety and Security

Types of Cruise Ship Injuries

Cruises are obviously meant to be relaxing and fun, but there are many potential hazards on cruise ships that can injure you. Highly regarded cruise lines make passenger safety their number one priority and put a lot of effort into accident prevention. When you have millions of customers on your ships per year, safety simply cannot be ignored. Luckily, rules and regulations enforced by the US Coast Guard and Safety of Life at Sea (SOLAS) have been established to help prevent injuries and fatal accidents at sea. However, there is still an abundance of injuries that occur on cruise ships, and there are many causes.

Slip and fall injuries are very common and occur on cruise ships just as easily as they do on land. There is any number of reasons that a passenger might fall; a carpet may be frayed and improperly secured to the floor, or a piece of debris or some other obstacle may be obstructing a stairwell or hallway. Something as plain as a piece of ice sitting on a step can do some serious damage, especially if an elderly person slips on it.

Objects falling from upper decks can cause considerable harm when you take into account how large these ships are; some vessels can have as many as eight ascending balconies. Anything could possibly fall from above, but most likely it will be a bag or purse, drink, or even a child's toy. Many accidents like these happen when alcohol enters the picture, and gravity becomes an accomplice to the crime because the higher the object's point of release, the greater the impact it will have when it collides with a person below.

Fire on a cruise ship can be extremely dangerous, which is why the crew stresses fire prevention to such an extent. There are a variety of reasons that fires start, specifically faulty equipment or solvents and debris that is ignited as a result of some sort of negligence. Passengers risk being trampled during a fire if the crew is unable to manage evacuation at the exits. If the fire is so intense that passengers must abandon ship, they can injure themselves as they hastily depart from the vessel to enter a life boat or, in extreme cases, as they plummet into the water. Passengers may suffer from burns if they are close enough to the flames, but most fire injuries happen during the escape.

Cruise ships are like tiny cities, so they have all the luxuries of recreational activities crammed into a restricted area. In the past, passengers have been struck by golf balls at the driving range. Unbelievably, there have been instances where passengers come into the line of fire because they cross paths with skeet shooters. It may seem slightly ridiculous that there are activities involving guns on the ship, but this gives you an idea of how enormous these ships are and how serious safety needs to be taken while aboard a cruise. You may not even be safe while attending a show, as one passenger learned the hard way when she was struck by an object thrown at her by one of the entertainers.

Accidents at the pool are by no means unusual. As with any pool, injuries can occur if people slip and fall on the deck, carelessly jump or dive in and land on someone, or, in the worst case scenario, drown. Passengers may also become ill due to contamination of the water or improper water filtration.

Working out at the gym is a popular activity among cruise passengers, but there are risks associated with it that the cruise line cannot be held accountable for; if a passenger chooses to exercise knowing full well that strenuous activity may be detrimental to their body, the crew is not responsible for stopping that person from participating. On the other hand, if a weight machine or treadmill was defective or malfunctioned, and a passenger was hurt, then they may have grounds to sue.

If the cruise ship's captain and crew make a navigational error that causes the vessel to veer off course, passengers may be injured if it collides with another boat or a stationary object. If passengers are forced to jump ship under the circumstances, the crew must keep passengers calm so that they can empty out in an orderly fashion. Errant navigation can sometimes cause the ship to tilt abruptly. Passengers can be injured in the process if loose objects are jolted or pool water rushes across the deck.

Severe weather conditions and natural disasters can create major problems on cruise ships. Hurricanes, earthquakes, and other phenomena can cause tidal waves, high winds, etc. that can wreak havoc on a ship. Sometimes these events are unpredictable and impossible to prevent, but there are occasions where the crew has been forewarned and fail to adjust the ship's course to get out of harm's way. Predicaments like these have even resulted in cruise ships sinking.

Passengers visiting the infirmary while onboard a ship can become the victims of medical malpractice at the hands of a doctor or nurse. A case like this can get complicated if the cruise ship is flying a foreign flag because the physicians on staff may not have earned US medical licenses. This does not necessarily mean that doctors trained in other countries are inferior, but problems can arise if you are a US citizen trying to prosecute the case due to the licensing discrepancy.

Another familiar source of cruise ship passenger injury is poisoning. Poisoning can originate from tainted food or viruses and airborne pathogens being spread through vents or heating and air conditioning units. Food poisoning often leads to gastroenteritis, which triggers bouts with nausea, and vomiting, diarrhea, and abdominal cramps.

Similar to gastroenteritis, E. coli is a strain of bacteria that contaminates food like beef and vegetables. Its effects are almost identical to gastroenteritis. Cruise ships can be havens for infection like legionnaire's disease, norovirus, and other forms of sickness based on the fact that there is such a dense population confined to a limited space. Legionnaire's disease is a bacterial infection that causes pneumonia, and norovirus is what causes the common stomach flu.

Since cruise ships are bound to make stops at various ports throughout the duration of the trip, there are numerous onshore activities that could ultimately be unsafe. These include:

- Tours
- Horseback rides
- Bungee jumping
- Helicopter rides
- Jet skiing
- Parasailing
- Water skiing
- Swimming.

Passengers should be aware that participation in many of these activities is at their own risk and has no affiliation with the cruise line.

Cruise Ship Sexual Assault

Although there are laws in place protecting victims of sexual assault and rape on cruise ships, police endeavors to enforce them have essentially been nonexistent. The only method this leaves to obtain any justice is pursuing a civil suit for damages incurred.

Research revealed that over a span of four years, out of 174 reports of sexual assault or rape on one of the most renowned cruise lines, not one person has been convicted of a crime. This could be due to the fact that this cruise line makes a concerted effort to conceal any evidence that horrific events like these are even happening on their ships.

According to this particular corporation's Risk Management Manual:

1. As all accidents or illnesses may result in litigation, the Risk Management Department should be promptly notified in the event of an accident or incident, to jointly review and analyze the information received from the ship's initial investigation, to determine the steps necessary to further investigate and defend any claims arising out of the accident/incident including immediate contact and retention of defence counsel.

2. As required, make arrangements to transport defence counsel investigators to next port of call.

What to do if a Sexual Assault has Occurred

The crime scene must not be tampered with so that all evidence is preserved. The following steps should be taken:

- The scene should be closed off from intruders until police get there.
- Photographs should be taken from all angles of the scene.
- No one should be allowed access to the scene other than officials capable of collecting evidence without ruining it.
- Anyone inside the scene should be wearing gloves and booties.
- Any eyewitnesses should be detained until their accounts of the event have been recorded and names, addresses, and telephone numbers have been obtained.
- The victim of the sex crime should not eat, drink, shower, bathe, brush their teeth, or go to the bathroom until a doctor conducts a rape test.
- Blood samples should be collected from the victim as soon as possible.
- Pictures of the victim should be taken to showcase the extent of harm (e.g., cuts, bruises, black eyes, swelling).
- Only qualified individuals should interact with the victim.
- Mug shots of the entire staff and passengers should be shown to the victim to identify the assailant.

You cannot depend on the cruise line to take action and abide by the above rules. Friends and family of the victim should take matters into their own hands and call the FBI by satellite phone to find out where to take it from there. Leaving it up to the cruise line will probably not get you very far. Aside from contacting the FBI, you should also get in touch with a lawyer who specializes in cruise ship sexual assault cases immediately.

Stunning Statistics

There have been some shocking revelations stemming from the investigation of various cruise ship rape cases.

- One leading cruise line averages two reports of sexual assault or rape for every 100,000 passengers. That's like saying two people in a packed football stadium will be raped.
- Another major cruise line documented 99 sexual assaults or rapes during a five-year window.
- A different company reported 173 sexual assaults or rapes over a five-year span.
- Another large brand gets approximately two reports of sexual assault or rape every month on each ship in its fleet.
- One cruise line employs a risk management team focused on sexual assault and rape claim investigations. It is confusing why a risk management team is assigned this task as opposed to the security department.

Frequently Asked Questions About Cozumel Health & Safety

Before we launch into answering the most frequently asked questions about health and safety on Cozumel island, we'd like to say a special word to our cruise passenger visitors.

It's common practice on the cruise ships for the information officers to tell you Cozumel is not safe. Unless, that is, you stay in the "approved areas."

What a coincidence that every single establishment on this safe list for shopping and eating kicks back a significant percentage of sales to the cruiseship company.

In short, if you believe what they tell you about how you will be cheated, how your pocket will be picked etc.if you stray back into town or off on your own or buy from an "unauthorized" store, you are falling into their trap.

And that's really a pity. Because Cozumel is a remarkable destination from the safety point of view.

Cozumel is, in fact, the very favourite stop of most crew members. Why? Because they can roam the streets of town into the wee hours of the morning without worrying for their safety.

And they can eat and shop a few blocks back off the waterfront cheaply and away from the crowds that pack the areas near the cruiseship pier and on the downtown waterfront.

We live in town — three blocks back from the waterfront and 4 blocks from the Punta Langosta cruiseship pier. Carey regularly goes out late at night, by herself, for ice cream. This is one of the things we value most about living on the island. It is unusually safe. So don't let the cruiseship companies put the fear in you for their own gain! Get out and experience this friendly little island — on your own terms!

And now we will, ahem, get off our soap box and get to the most frequently asked questions.

Is it Safe to Drink the Water?

Although Cozumel's water supply is chlorinated to some extent, you should never drink the tapwater. (This includes anywhere in the Yucatan, by the way—even if you are staying at a major resort with it's own filtration system.) Vacation rentals provide bottled water in 5 gallon water cooler jugs and some hotels provide small bottles each day. You should drink that. In addition, this water is made via osmosis and tastes great and isn't salty and hard like the tapwater.

How Safe is it to Eat in Local Restaurants?

All the restaurants recommended on this site use bottled water for food prep, drinking and ice. Although we have not yet had time to review every good restaurant in Cozumel, the ones you'll see listed on this site continue to pass our stringent standards for cleanliness.

We would, however, urge you not to buy food from street vendors.

I Want to Eat the Local Produce. Is It Dangerous to Do This?

The selection of vegtables and fruit available on Cozumel is outstanding and so are the prices. If you like pineapples, mangos, papayas and so much more, you can enjoy it all you wish on your trip. Just follow these simple steps before you eat it.

1. Purchase a small bottle of an iodine-based water purification drops. These are readily available in the

produce sections of the island's three big groceries. They're usually hanging up.

2. Fill the sink or a large pan with tap water and add the drops to it. A "gota" is a drop. A liter is around a quart and that's all you need to estimate how many drops for the amount of water you're using.
3. Pre rinse your produce if it looks gritty and then dump it in the water bucket to soak for 20 minutes or so. So this before you even put it in the refridge. And we would advise doing to all produce including bananas.

By the way, when you eat unwashed produce in the U.S. you run some of the same risk, although in the U.S. produce is washed (well, usually) and when YOU wash it in the U.S. you are using safe water (well, usually). So after you soak the produce as described above, it is safer than what you often get at home. We have NEVER been sick from consuming locally-bought raw produce that was first treated in the manner described. Try it, you'll like it.

Are There Any Other Measures I Should Take to Avoid Becoming Ill in Mexico?

The most important thing you can do to keep yourself healthy is wash your hands a lot. Wash them every time you handle money, for example, after you've pushed a car through the grocery store, etc. etc. It makes good sense and it can make all the difference. Some people swear by chewing pepto bismal tablets morning noon and night. We think this isn't necessary.

We do suggest the following preparations for the unlikely event of becoming ill. Bring a small supply of Immodium-AD. This will not cure you. But it will stop stomach cramps and diarhea while you can get to the heart of the problem. Ask your doctor back home what antibiotic and dosage he/she would recommend in the event that you come down with Monteczuma's Revenge if it doesn't disappear on its own in about 24 hours on its own. Write down the generic name and bring this information with you.

Ciprofloxin and other modern antibiotics are readily available here, without a prescription and at a quarter or more the US price at every pharmacy on the island. This includes the large Chedraui and St. Francis D'Assis grocery stores which each have a pharmacy in them.

If you have no Spanish and feel you can't communicate, hit the American DrugStore on Calle 3 two doors in from the waterfront downtown. They have everything and speak great English.

Unlike some places in Mexico AND the United States, incidents of violent crime are a fairly remarkable occurrence here on the island. Cozumel's small city, San Miguel, population approximately 80,000, is full of Mexican middle-class folks who moved from other parts of Mexico for that very reason. They vehemently believe that the island is the safest place in the country to raise their children.

It's partly the culture—conservative, deeply religious with a very strong Mayan component. But it's also the fact that this is an island and isolated from the mainland. So the growing crime statistics we read about for Cancun and the Yucatan coast don't impact much here. You have to realize that it costs more than a typical day's wages to take the ferry over here and another day's work to get back. This definitely keeps the riff-raff down.

Also the police here are vigilant — both the special tourist police and the ones that work back in the neighbourhoods of San Miguel. Cozumel is one of the brightest crown-jewells in the Mexican tourist economy. The powers that be want to keep it clean and keep it safe. And in our book they're doing a remarkably good job of just that.

This is not to say there's no crime at all, of course. But 99.9% of it is property—break-ins to houses left obviously empty for too long, cameras stolen from rental cars that weren't locked, that kind of thing.

Some in the states have called us overprotective parents. Yet in San Miguel we feel very comfortable letting our 20-year-old daughter venture out at night with her girl friends. After a lifetime of having to be paranoid about walking alone at night in cities, Carey says this is one of the things she most appreciates about Cozumel. And safety-wise it is torso, shoulders and head above other places we've visited in the Caribbean including St. John and St. Thomas, VI.

So use common sense and stay alert walking at night. Because there's a few rotten apples in every barrel. But we seriously doubt you'll run into any of them on Cozumel island.

What Kind of Medical Service is Available Locally?

There is excellent medical service on the island now and your insurance will cover you for problems. You also will have the novel experience of Not Having to Wait Very Long. We like the San Miguel Clinic on Calle 6 between Avenida 5 and 10—a block back from the waterfront Cozumel Musuem.

We had a guest last month who had a recurrence of a kidney infection while visiting. We whisked her in there and they had her in an observation room with a nurse taking her blood pressure in a matter of about 5 minutes. Within 10 minutes a doctor who spoke beautiful English was at her side. They diagnosed the problem, kept her overnight, fixed her up and sent her home. The parents told me that their insurance covered all but $100 of the expenses.

Another good-sized well-equipped facilitiy with many capable English-speaking doctors is the Cozumel Medical Centre (CMC) which is on Calle 1 between Avenida 50 and 55.

If you have a problem while diving, there are two hyperbaric chambers on the island.

lifetime.

If You Take Regular Medication

Ensure that you have sufficient supplies to last the length of your holiday. It is also wise to consider taking a couple of days' extra supply in case of unforeseen changes in your plans. Remember to always carry your medication in your hand luggage. If you have recently received treatment for any significant medical condition, it is suggested that you carry a recent report with any important information from your doctor.

Motion Sickness

If you are concerned, you should purchase medication before you leave home. It is best to take the tablets just before sailing to prevent illness. Seasickness medications are also available at a charge on board.

Be Safe in the Sun

Avoid excessive exposure to the sun between 10.00am and 3.00pm. Remember 'Slip! Slop! Slap! Wrap!' – Slip on a t-shirt, Slop on high factor sunscreen, Slap on a hat (wide-brimmed is best so that your ears are protected too), Wrap on sunglasses. Wear a t-shirt to protect yourself from the sun while swimming. Re-apply sunscreen frequently after swimming or exercise.

Drink in Moderation

When you are ashore, be aware of the alcohol content of local and imported drinks. You are much more likely to suffer a serious injury if you are under the influence of even small amounts of alcohol.

Be extremely careful if you drink Kava which is a popular herbal, medicinal and ceremonial drink in the South Pacific. In addition to enhancing the effects of alcohol, long-term use may cause liver damage.

Avoid Biting Insects

Wear light coloured, loose fitting clothing that covers your arms and legs and regularly apply an insect repellent that contains at least 20% DEET.

Wash Your Hands

To help prevent illness, wash hands often and thoroughly and always after going to the toilet, before handling food and eating.

Pregnancy

We are unable to take on board any passengers who will have entered their 24th week (or later) of pregnancy by the end of the cruise.

The Medical Centre

Should you experience any medical issues during your cruise, each ship has a medical centre staffed by registered doctors and nurses, available during scheduled clinic times. They are also available 24 hours a day, 7 days a week for medical emergencies. Our medical centre is only for medical needs arising on board and cannot cater for treatments that you know you will require while on holiday. We are not a healthcare provider and may not be held reliable for the sickness, injury or death of any passenger arising from any advice, treatment, care, services or any omission by medical staff. These medical providers exercise their own medical judgement and expertise.

All cruises on our ships are outside the scope of Australian Medicare, New Zealand "ACC" and private health insurance. Consultations, treatments and medication are charged at private rates and must be paid by you and claimed through your travel insurance.

Travel Insurance

We strongly recommend that you take out comprehensive international travel insurance at the time you pay your deposit. Ensure your insurance covers you for general medical expenses, existing medical conditions, hospitalisations and medical evacuations.

If you plan to participate in sports such as scuba diving or parasailing...ensure your policy covers these activities.

If you are going on an Australian coastal cruise...you still require international travel insurance, as your cruise is outside the scope of Australian Medicare, the New Zealand Accident Compensation Corporation (ACC) and private health insurance.

Be aware when organising your insurance that uninsured travellers (or their families) are held personally liable for cancellation, medical and associated costs incurred at any time whether before the cruise, on board or overseas. Hospitalisation costs on board or outside of the country and medical evacuation costs are very expensive and are not covered by Medicare, ACC or private health insurance. Daily hospital bills can cost as much as AUD5,000 per day. All consultations, treatments and medications are charged at private rates.

Vaccinations

Don't leave these till the last minute. Wherever you're travelling we recommend you have a minimum of Influenza (seasonal), Hepatitis A and Tetanus/Diphtheria vaccinations. We also strongly advise you to seek advice from either your doctor or an approved public health travel advisory service to identify any specific vaccinations or health precautions required for each port of call. E.g.

- World Health Organisation
- The Department of Foreign Affairs and Trade in Australia
- Smart Traveller
- New Zealand-New Zealand Ministry of Foreign Affairs & Trade.

Some countries strictly enforce certain health requirements. If you don't have correct vaccinations, you may not be able to go ashore at certain destinations. Additionally, some local authorities may impose a monetary fine or deny the ship entry into the port.

Special Diets

If you have special dietary requirements, we ask that you confirm your dietary requests with the Maitre d' or Head Waiter upon arrival. They will discuss the details with you to avoid any misunderstanding.

Please note the Doctor onboard advises against eating at the buffet due to the risk of cross contamination of foods by passengers moving food from one container to another, thereby possibly dropping contents into another bowl.

Generally, you will be requested to choose an item from the usual dinner menu. These will be available one night in advance for your information (breakfast and luncheon menus are also available in advance upon request). If there are any special requirements, we request that you make this known to the Head Waiter, or Maitre d'Hotel the evening prior to the meal in question, preferably during dinner.

While we will endeavour to meet your specific needs and requirements, P&O Cruises ask that you take responsibility for your own needs by taking the extra time to peruse the menus and make your selections accordingly.

Any concerns or questions regarding meals and diet can be addressed to either the Maitre d'Hotel or Head Waiter who will do their utmost to assist in any way they can.

Accessibility

P&O Cruises will make every effort to accommodate passengers with disabilities. We must be notified of wheelchair usage and/or any special needs at the time of booking. We have wheelchair-accessible cabins on all P&O Cruises ships, subject to availability at the time of request. We also have specialised kits for hearing impaired passengers available, upon request at time of booking. If you require a mobility device, you must provide your own. If you have any special requirements or questions regarding mobility, please inform your travel agent, or contact our Telephone Sales & Service Team. When bringing your own wheelchair, we highly recommend collapsibles wheelchair, as the width of cabin doors may vary. Some P&O Cruises ships have areas that are not wheelchair accessible.

Ports Accessed Via Ship's Tender

It is important to note that access to tender ports may be restricted and this may limit your tour options or restrict access to these ports. Do not worry if you cannot get off the ship, we still have plenty of fun activities for you on board!

Shore Tours

Many of our shore tours are suitable for people with limited mobility. However, access to tender ports may be restricted. Our shore tours staff will happily suggest suitable tour itineraries to accommodate your mobility needs. Please note any additional costs incurred are the responsibility of each passenger.

Security on Board

At P&O Cruises, we want you to have a safe and enjoyable cruise. As a result, we take the safety and security of our passengers and crew very seriously. Closed Circuit TV (CCTV) monitoring has been added across our fleet to enhance on-board security. We continually review the security situations at every port we visit. We have also introduced additional screening in ports, including the use of drug sniffer dogs at embarkation and random itemiser checks at ports of call.

Never accept drinks from strangers and never leave your drink unattended. If you think your drink has been spiked, report this immediately to either the local authorities if ashore, or our security staff or bar staff on board.

Dangerous Articles/Illicit Substances

All luggage will be security screened for drugs and alcohol as well as dangerous weapons at embarkation. P&O Cruises reserves the right to deny boarding or to disembark any persons in possession of illicit substances.

Dangerous articles such as fireworks, guns or other weapons and flammable liquids, etc, are strictly prohibited and must not be included in baggage brought on board nor kept in cabins under any circumstances. It is an offence to carry such items. Any offence committed by passengers will be reported to the appropriate authorities.

Alcohol Policy

Passengers under 18 years of age are not permitted to purchase or consume alcohol on board. Government issued photo identification (Queensland proof of age cards will not be accepted) will be requested. We wish to advise that passengers are not permitted to take on any alcohol for personal consumption.

Duty-free and tax-free and personal alcohol purchased on-shore, must be surrendered by passengers at embarkation or prior to boarding in any port. Duty and tax-free alcohol is also available for purchase on board and will be held under bond and returned at disembarkation in the final port of call. Please note, all luggage, including hand and check-in will be x-rayed at embarkation and disembarkation. Individuals must declare to Customs all amounts of alcohol in excess of the local Customs limits.

Smoking Policy

In the interest of our passengers' health, safety and enjoyment during their cruising holiday, smoking is not permitted indoors, including in cabins, suites and on private balconies. This is in line with Australian standards. Outdoor smoking venues will be published in the Pacific Daily on board your cruise.

Right to Deny Boarding

P&O Cruises reserves the right to deny boarding of any person it believes is not suitable to travel. This could be because the person:

- Is unfit to travel in the opinion of the Senior Doctor
- Seeks to travel in violation of any applicable law
- Has a serious criminal conviction and/or in the reasonable opinion of P&O
- Cruises, poses a significant risk to other passengers and/ or crew
- Breaches the applicable Terms and Conditions
- Possesses, or is under the influence, of drugs or alcohol
- May compromise the safety or comfort of other passengers or for the protection of property
- Uses offensive language or behaviour towards fellow passengers or crew.

Safety and Security on Shore

A few helpful hints to help you enjoy yourself in safety onshore.

- Be security conscious and take sensible precautions. Always pack your own luggage and don't leave it unattended.
- Obey the local law – remember that the laws and procedures which apply are those of the country you are in, not your home country.
- Dress and behave conservatively, in accordance with local customs and sensitivities.
- Always swim at safe beaches and only participate in snorkelling or scuba activities if you have received training. Do not scuba within 24 hours of flying. Never dive into water of unknown depth.
- Do not swim after drinking alcohol.
- When going ashore carry only the minimum amount of cash that you will need for the day. Leave your valuables and spare cash and credit cards in your cabin's safety deposit box.
- You must wear a life jacket for boating, water skiing and other offshore water sports.

- Check water sports equipment for quality and safety features.
- Booking tours outside of the P&O Cruises Island Discoveries programme is not recommended as P&O Cruises has not individually assessed tours outside this programme and cannot make any comment on the standard or safety of these tours.
- Passengers are not permitted to take food of any kind from the ship to any destination visited. During your time ashore, we recommend that you drink plenty of bottled water, avoid ice and eat well-cooked food only. Beware of uncooked food including salads and fruit that you cannot peel. The conditions and method of preparation of food and drinks cannot be monitored by P&O Cruises. With that in mind, passengers are advised that if they purchase or consume food at their visiting ports, they do so entirely at their own risk.

Travelling with Children and Teenagers

To ensure there is suitable supervision, there needs to be at least one passenger 18 years or older in each cabin. However, to accommodate families cruising together, children may occupy a separate cabin to their parent/guardian, providing one child is aged 16 years or over. For cruises departing from an Australian port between, between 1 Nov and 7 January, any passenger under 19 years of age on the day they board they cruise, must travel in the same cabin as a responsible adult 19 years or older. In addition from 8 January to 31 January there is a limit on the number of passengers under 19 years of age who can travel unaccompanied. Once the limit is reached, a responsible adult 19 years or over must travel in the same cabin. We can advise you at the time of booking whether the limit applies to your booking. The responsible adult is accountable for passengers under 19 years of age in their cabin for the duration of the cruise, and the 'Rights of the Captain' will be applied to anyone who displays disruptive behaviour.

We welcome children 12 months and over to cruise with us. We are unable to accommodate infants under 12 months due to limited neo-natal facilities onboard and at ports of call. In addition, children under three years must remain with their parent/legal guardian at all times.

Children with Special Needs

P&O Cruises welcomes all children and teens aged 3 years to 17 years to our fun programmes. We need to be advised at time of booking or at least 60 days prior to sailing if any children have medical and/or special needs, which we make every effort to accommodate. Please be advised that P&O Cruises youth staff do not offer individual one-on-one supervision. If your child requires one-on-one supervision, they will need to be accompanied by a parent / legal guardian at all times.

Children with Allergies

If your child suffers from life-threatening allergies, eg. to nuts, they are welcome in our Youth Centres. However, Youth Centre staff members are unable to take responsibility for administering the Epi-Pen or any other medication.

We will always advise you if food is to be served within the childcare area.

In the event of an emergency, staff will follow the standard on-board medical emergency protocol, which involves the qualified doctors and nurses on your ship.

Cruise Cards

As you embark, you'll receive a personalized Cruise Card. All minors are issued with a red Cruise Card and passengers over 18 receive a blue Cruise Card. This way our staff can always identify children aged under 18 years as minors.

It is the responsibility of parents/legal guardians to monitor their child's expenditure during the cruise.

Restrictions to Gambling and Alcohol Cconsumption

In line with domestic Australian age limits, as well as P&O Cruises policy, the casino and cash prize bingo are strictly reserved for passengers 18 years and over. Additionally, only those passengers 18 years and over will be allowed to purchase or consume alcohol. Government issued photo identification (Queensland Proof of Age cards will not be accepted) may be requested. Please be aware that if an adult attempts to buy alcohol for anyone under the age of 18, both the adult and minor will be reported. A range of actions may be taken, including disembarking those involved.

Family Information Pack

For more information on travelling with children, we have a comprehensive family information booklet available. This pack will provide you with all the information you require if travelling with children.

Upper Bunk Beds

The safety of your children is very important to us at P&O Cruises. In line with Australian Standards it is strongly recommended that children under the age of 9 years do not play or sleep on upper bunk beds.

Public Areas

Children under 13 years must be accompanied by a parent/legal guardian in the public lounges after 9.30pm. Be sure to check with the Youth Centres for special evening events designed for kids and teens. Children under the age of 18 years must be accompanied by a parent /legal guardian in the entertainment lounges and restrictions may apply to access to pools, whirlpool spas, gym and Lotus Day Spa.

Children and Pool Activities

Pool activities are not included in our youth or teen programmes. Any child using the pools on board must be supervised by a parent or guardian at all times. For public safety reasons, children in nappies and children who are not toilet trained are not permitted in the public swimming pools and whirlpool spas on board.

Travelling with Infants & Toddlers

We welcome children aged 12 months and over to cruise with us. We are unable to accommodate infants less than 12 months of age due to limited neo-natal facilities on board and at ports of call. Infants aged between 12 months and 3 years must remain with their parent/guardian at all times.

Cots

We strongly recommend that children aged 12 months to 2 years 11 months sleep in a cot. Cots are permitted in all grades of accommodation, providing that the number of passengers in the cabin does not exceed the maximum capacity (ie. 3 people in a 3 berth cabin or 4 people in a 4 berth cabin).

When a cot is being used, a queen bed cannot be accommodated, except in a suite or mini-suite.

Baby Baths

Unfortunately, these are not available on any P&O Cruises ships.

Nappies and Baby Wipes

Parents or guardians need to provide all of their own nappies, wipes and other necessary baby accessories for the length of the cruise. These items are not available on board.

Prams/Strollers

Unfortunately these are not available on board. Please ensure that you bring your own.

High Chairs and Booster Seats

These are available at on-board dining facilities on request.

Food for Infants

Baby food and baby formula are not supplied on board our ships. However, we are able to store any food or formula carried on board. If you need to store any food during your cruise please contact Customer Care. Please ensure you travel with enough food to cover the entire cruise and a little extra for any unforeseen delays.

Illness on Cruise Ships

In 2004, 10.8 million people took North American cruise vacations, and cruising is expected to continue to gain popularity, with an estimated 20.7 million cruise travellers in 2010. Since 1980, the North American cruise industry, which makes up the majority of the global cruise market, has experienced an average annual passenger growth rate of 8.1%. Cruise capacity also increased by 450% over the same period, with 225,714 beds available in 2004. U.S. ports handle about 75% of global embarkations, and 77% of the passengers are U.S. residents. The North American cruise industry has had overall passenger occupancies at full capacity in recent years, with top destinations being the Caribbean, the Mediterranean, Alaska, Europe, and the west coast of Mexico. Cruise itineraries are ever expanding to include continents and areas not easily accessible by other means of travel, permitting travellers to visit multiple global destinations in the comfort typically offered by cruise ships.

Cruise ships and all ocean-going vessels engaged in international commerce fly flags of registry, which are required for operation in international waters. Cruise lines often choose to register either in their country of ownership or ship production or in countries that provide open maritime registration, or "Flags of Convenience." Ships are most often registered in the United Kingdom, Liberia, Panama, Norway, the Netherlands, the Bahamas, and the United States.

Flag registry states provide comprehensive maritime expertise and administrative services, require annual safety inspections before issuance of a passenger vessel certificate, and monitor vessel compliance with international maritime laws, as well as flag state standards. The World Health Organization's International Health Regulations stipulate health and sanitation requirements for international conveyances. In the United States, the U.S. Coast Guard enforces maritime safety requirements, and CDC has regulatory responsibilities for sanitation and public health on cruise ships bound for a U.S. port from a foreign port.

Today's large cruise ships can serve as a gathering place for the global community, where passengers and crew from around the world bring together a diversity of cultures, as well as medical and immunization backgrounds and health risk behaviours. Cruise voyages can last from several hours (e.g., gambling cruises) to several months (e.g., around-the-world and semester-at-sea cruises); however, the average duration of cruise travel is about 7 days. This time period permits ample opportunity for passengers and crew to come into repeated and prolonged contact through shared activities, such as games and dining, and through resources such as food and water, resulting in opportunities for exposure and transmission of communicable diseases. In addition, as cruise ships make multiple port stops, where differences may exist in sanitation standards and disease exposure risks, embarking passengers and crew can import and spread communicable diseases onboard. Detecting and preventing infectious diseases acquired during cruises are important not only to protect the health of cruise travellers but to avoid global dissemination of diseases in home communities through disembarking passengers and crew members.

Cruise Ship Medical Facilities

In 2000, the Cruise Ship and Maritime Medicine Section of the American College of Emergency Physicians (ACEP) published ACEP Health Care Guidelines on Cruise Ship Medical Facilities, a consensus report on appropriate facilities and staffing able to provide basic medical and emergency services aboard cruise ships, within the recognized limitations of the offshore environment. These guidelines include provisions for medical facility design on cruise ships, including guidelines for an isolation room to manage communicable diseases, diagnostic and emergency medical equipment, formulary, staff number and qualifications, and a health, hygiene and safety program for medical personnel.

Large cruise lines that operate in the United States or are members of the International Council for Cruise Lines (ICCL) meet or exceed the ACEP guideline standards. Medical facilities on ICCL member vessels can be equated to community urgent care centres, with the ability to perform basic diagnostics such as blood chemistries, complete blood counts, urinalyses, chest x-rays, and EKGs. An estimated 95% of illnesses seen in cruise ship medical facilities can be treated onboard; however, passengers with serious problems such as myocardial infarction or cerebrovascular accidents need to be transferred to shoreside hospitals after stabilization. Cruise travellers should note that ACEP guidelines for large cruise lines may not be followed by smaller ships or those run by independent operators; on such ships there may be no medical provisions onboard. Cruise ship travellers with chronic diseases or those who may require comprehensive medical care during travel should consult with their health-care providers and notify the cruise line of special needs before travel.

Preventive Health for Cruise Ship Travellers

Due to multiple port visits and potential exposures, cruise ship travellers may be uncertain about which prevention medications, immunizations and behaviours are appropriate for them and for their itineraries. In general, travellers should inform the cruise line of special medical needs, such as wheelchair access, oxygen tank, dialysis, etc, in advance of travel. Adequate medical insurance coverage for receiving health care overseas and medical evacuation should be ensured. Four to six weeks before travel, cruise travellers should consult a health-care provider who can give guidance on appropriate chemoprophylaxis, immunizations, and health behaviours based on a complete review of the health status of the traveler, duration of travel, countries to be visited, and shoreside activities. Clinicians should provide travellers with

1) destination-specific recommended and *required vaccines* (e.g., yellow fever), as well as prevention medication (e.g. malaria chemoprophylaxis) if needed;
2) routinely recommended age-and medical condition-*specific immunizations*, such as MMR and influenza vaccines, if not up to-date;
3) appropriate options for *motion sickness* based on the individual's medical history and current medications and
4) a *written summary of their medical history*, including pertinent diagnostic data such as EKG and chest x-ray, to facilitate overseas medical care, should it be required. Among cruise ship passengers and crew members, risk of exposure to infectious diseases is difficult to quantify because of the broad spectrum of cruise ship experiences and limited data. Travellers should be given guidance on good health habits and disease prevention practices during travel including adequate.
5) *hand hygiene* (i.e., washing hands for at least 20 seconds with soap and water) or if soap and water are unavailable, using an alcohol-based product containing more than 60% alcohol

6) *respiratory hygiene* such as using tissue to cover coughs and sneezes;
7) *food and water intake precautions* (i.e., eating foods that are thoroughly cooked and of appropriate temperate); and
8) *mosquito prevention* (i.e., using DEET-containing repellents, bed nets, and clothing that provides more coverage over exposed areas of the body. Health-care providers can contribute to healthy environments on cruise ships by questioning ill returned travellers about recent cruise vacations and promptly reporting any suspected communicable disease to public health authorities.

Handwashing on Cruise Ships

To stay clean and healthy, wash your hands with warm water and soap.

When to wash your hands:

Before:

- Touching your hand to your mouth including:
  - o Eating and drinking,
  - o Smoking,
  - o Brushing your teeth,
- Helping a sick person.

After:

- Going to the bathroom,
- Changing diapers,
- Touching high-hand contact surfaces, such as:
  - o Door knobs,
  - o Elevator buttons,
  - o Railings,
- Returning to your cabin,
- Helping a sick person, and
- Blowing your nose.

How to wash your hands:

1. Wet your hands with warm water.

2. Apply a generous amount of soap.
3. Rub hands together for 20 seconds.
4. Rinse hands.
5. Dry hands with paper towel.
6. Use paper towel to turn off the faucet and open the door.

What about alcohol-based hand sanitizers?

- CDC recommends that cruise ship passengers use warm water and soap to wash their hands. Washing is always best.
- If water and soap are NOT available (perhaps on excursions), use an alcohol-based hand sanitizer with 62% ethanol based hand-sanitizer, preferably in a gel form.

CDC's Vessel Sanitation Program (VSP) and Quarantine Stations

In 1975, in response to several large gastrointestinal disease outbreaks on cruise ships, CDC established the VSP, a joint cooperative program with the cruise industry, to achieve a high level of sanitation and minimize the risk of gastrointestinal (GI) disease on cruise ships. The VSP encourages the cruise industry to establish and maintain a comprehensive sanitation program, including surveillance for acute gastroenteritis (AGE). The VSP staff conducts biannual, unannounced sanitation inspections on U.S.-bound cruise ships with international itineraries carrying 13 or more passengers. The VSP also engages in the design and construction of new ships, as well as retrofitting older ones to enhance facilities and provisions that promote shipboard sanitation and public health.

The VSP shipboard sanitation inspections cover the following areas of public health interest:

1) water supply storage, distribution, disinfection and protection;
2) spas and pools disinfection and filtration;
3) food handling, including storage, preparation, and service;
4) potential for contamination of food, water and ice;
5) personal hygiene and sanitation practices of crew;

6) general cleanliness and condition of the ship, and
7) ship training programs in environmental and public health practices.

An inspection score of 86 or higher (out of 100) indicates an acceptable level of sanitation. In general, the higher the score, the higher the level of sanitation, but this score does not reflect the risk of acquiring gastrointestinal disease. The VSP could recommend or require that a cruise ship not sail if sanitation deficiencies could pose a public health threat.

U.S. Federal Quarantine Regulations require that, 24 hours before arrival at a U.S port, vessels notify public health authorities of onboard incidents of death, diarrhea (defined as at least 3 stools in a 24-hour period), and certain febrile syndromes of public health significance. The VSP receives reports of AGE and may respond with an epidemiologic and environmental investigation if at least 3% of passengers and/or crew members seek medical attention for AGE, or if an unusual GI illness occurs. Surveillance and response for reports of onboard deaths and febrile syndromes of public health concern are conducted by CDC Quarantine Stations, located at major U.S. ports of entry and land border crossings and administered by the Division of Global Migration and Quarantine in Atlanta.

Below are the current ships that have scored 100% on the last report. They are checked at least two times a year to keep the standard high on the cruise lines.

What is Viral Gastroenteritis?

Gastroenteritis means inflammation of the stomach and small and large intestines. Viral gastroenteritis is an infection caused by a variety of viruses that results in vomiting or diarrhea. It is often called the "stomach flu," although it is not caused by the influenza viruses.

What Causes Viral Gastroenteritis?

Many different viruses can cause gastroenteritis, including rotaviruses, noroviruses, adenoviruses, type 40 or 41, sapoviruses, and astroviruses. Viral gastroenteritis is not caused by bacteria (such as Salmonella or Escherichia coli) or parasites (such as

Giardia), or by medications or other medical conditions, although the symptoms may be similar. Your doctor can determine if the diarrhea is caused by a virus or by something else.

What are the Symptoms of Viral Gastroenteritis?

The main symptoms of viral gastroenteritis are watery diarrhea and vomiting. The affected person may also have headache, fever, and abdominal cramps ("stomach ache"). In general, the symptoms begin 1 to 2 days following infection with a virus that causes gastroenteritis and may last for 1 to 10 days, depending on which virus causes the illness.

Is Viral Gastroenteritis a Serious Illness?

For most people, it is not. People who get viral gastroenteritis almost always recover completely without any long-term problems. Gastroenteritis is a serious illness, however, for persons who are unable to drink enough fluids to replace what they lose through vomiting or diarrhea. Infants, young children, and persons who are unable to care for themselves, such as the disabled or elderly, are at risk for dehydration from loss of fluids. Immune compromised persons are at risk for dehydration because they may get a more serious illness, with greater vomiting or diarrhea. They may need to be hospitalized for treatment to correct or prevent dehydration.

Is the Illness Contagious? How are these Viruses Spread?

Yes, viral gastroenteritis is contagious. The viruses that cause gastroenteritis are spread through close contact with infected persons (for example, by sharing food, water, or eating utensils). Individuals may also become infected by eating or drinking contaminated foods or beverages.

How does Food get Contaminated by Gastroenteritis Viruses?

Food may be contaminated by food preparers or handlers who have viral gastroenteritis, especially if they do not wash their hands regularly after using the bathroom. Shellfish may be contaminated by sewage, and persons who eat raw or undercooked shellfish harvested from contaminated waters may get diarrhea. Drinking water can also be contaminated by sewage and be a source of spread of these viruses.

Where and when does Viral Gastroenteritis Occur?

Viral gastroenteritis affects people in all parts of the world. Each virus has its own seasonal activity. For example, in the United States, rotavirus and astrovirus infections occur during the cooler months of the year (October to April), whereas adenovirus infections occur throughout the year. Norovirus outbreaks can occur in institutional settings, such as schools, child care facilities, and nursing homes, and can occur in other group settings, such as banquet halls, cruise ships, dormitories, and campgrounds.

Who gets Viral Gastroenteritis?

Anyone can get it. Viral gastroenteritis occurs in people of all ages and backgrounds. However, some viruses tend to cause diarrheal disease primarily among people in specific age groups. Rotavirus and norovirus infections are the most common cause of diarrhea in infants and young children under 5 years old. Adenoviruses and astroviruses cause diarrhea mostly in young children, but older children and adults can also be affected. Norwalk and Noroviruses are more likely to cause diarrhea in older children and adults.

How is Viral Gastroenteritis Diagnosed?

Generally, viral gastroenteritis is diagnosed by a physician on the basis of the symptoms and medical examination of the patient. Rotavirus infection can be diagnosed by laboratory testing of a stool specimen. Tests to detect other viruses that cause gastroenteritis are not in routine use, but the viral gastroenteritis unit at CDC can assist with special analysis upon request.

How is Viral Gastroenteritis Treated?

The most important of treating viral gastroenteritis in children and adults is to prevent severe loss of fluids (dehydration). This treatment should begin at home. Your physician may give you specific instructions about what kinds of fluid to give. CDC recommends that families with infants and young children keep a supply of oral rehydration solution (ORS) at home at all times and use the solution when diarrhea first occurs in the child. ORS is available at pharmacies without a prescription. Follow the written directions on the ORS package, and use clean or boiled water.

Medications, including antibiotics (which have no effect on viruses) and other treatments, should be avoided unless specifically recommended by a physician.

Can Viral Gastroenteritis be Prevented?

Persons can reduce their chance of getting infected by frequent handwashing, prompt disinfection of contaminated surfaces with household chlorine bleach-based cleaners, and prompt washing of soiled articles of clothing. If food or water is thought to be contaminated, it should be avoided.

Is there a Vaccine for Viral Gastroenteritis?

Currently there is a licensed rotavirus vaccine available that protects against severe diarrhea from rotavirus infection in infants and young children. Studies into a norovirus vaccine are underway.

Cruise Ship Safety

With the continued increase in the popularity of cruise vacations, there has been a similar increased in media attention on cruising. A vast majority of this media is extremely positive and focuses on the great value and new developments of today's cruise vacation. However, occasional news stories regarding ship safety, security and sanitation do occur which may generate questions from our clients and prospects.

Because safety — of both passengers and crew — continues to be the cruise industry's number one priority, the Cruise Line Coalition offers the following points to help respond to customers' potential questions and concerns regarding shipboard safety.

Note: This information is presented by the Cruise Line Coalition and not Cruise Direct Online.

Safety Standards

- All cruise ships must meet standards set by the International Maritime Organization (IMO) and the International Convention for Safety of Life at Sea (SOLAS).
- Ships operating from U.S. ports are also subject to U.S. federal and state regulations. The U.S. Coast Guard inspects all ships sailing out of U.S. ports four times a year.

- The International Council of Cruise Lines (ICCL) sets guidelines and mandatory standards for cruise companies seeking or maintaining membership in the association. These standards meet or exceed international and U.S. laws and regulations that apply to cruise ships.

Safety Equipment

- The average cruise ship has more than 4,000 smoke detectors;,
  - o More than 500 fire extinguishers;
  - o More than 6 miles of fire fighting hose;
  - o 5,000 sprinkler heads,400 fire stations or hydrants and sufficient lifeboats and life rafts for more than the number of individuals onboard.
- In addition, The the average cruise ship has more than 170 fully trained fire-fighting personnel, and many with advanced firefighting training.,
- Emergency and medical crews are only a few hundred feet from any possible incident. This means that the average response time in an emergency is a matter of minutes.

Onboard Security

- A cruise ship is comparable to a secure building with a 24-hour security guard. It is a controlled environment with limited access.
- Once underway, only documented employees and fare-paying passengers are permitted onboard.
- In addition, highly trained security personnel are employed on every vessel.
- U.S. laws protect American cruise line passengers. The FBI and federal authorities investigate and prosecute alleged crimes involving Americans in international waters.
- According to FBI statistics, cruise ships are safer than virtually anywhere in the U.S. in terms of crimes of any type.

Sanitation

- Sanitation standards on cruise ships far exceed those of any land-based facility.

- The industry and the U.S. Public Health Service's Centres for Disease Control and Prevention (CDC) have jointly established construction guidelines and operational procedures to maintain those high standards.

The cruise industry voluntarily works with the CDC, which oversees health and sanitary conditions on all passenger vessels visiting U.S. ports.

- Ships are subject to unannounced inspections by CDC officials, concentrating on fresh and drinking water, spa and pool systems, food storage, preparation and service areas, general hygiene, and waste management equipment and facilities.
- Ships are also periodically rated through the CDC's Vessel Sanitation Program (VSP) and must meet industry standards.

Protecting your Health while Aboard a Cruise Ship

When it comes to cruise ship safety, there are many individuals who are concerned with their physical safety. Physically staying safe aboard a cruise ship is important, but so is staying healthy. Due to a large number of individuals in a fairly small space, cruise ships have a number of health concerns.

The health problems and concerns found on cruise ships are often highly publicized. Despite this publicity, most cruise ships are fun, safe, and clean. Being concerned with your health aboard a cruise ship is one thing, but you should not let it prevent you from enjoying your vacation. Rather than letting the fear of getting sick ruin your vacation, you should take a few simple steps to ensure that your health will be the same as, if not better, when you first boarded.

Before your cruise ship is scheduled to set sail, you are urged to schedule a visit with your family physician. Be sure to inform him or her that you are planning a vacation aboard a cruise ship. Your physician will likely check your health to make sure you are fit for travel. In addition to examining your general health, it is possible that your physician may require a number of immunizations. These immunizations may offer you extra protection; therefore, you are urged to obtain them.

If you are taking multiple medications at the time of your cruise, you will want to make sure that you have an adequate supply. If you are running low on your medication, you may want to obtain a refill before you leave for your cruise. It is also likely that your doctor may request that you bring along extra medication. This request is often made to individuals who must take their medications to survive. In the event that you misplace your medication, you will have an extra supply to fall back on.

Taking health precautions before you board the cruise ship is important. However, it is also important to take precautionary measures after you board. These measures are simple; however, they may be vital to your health. One of the first things that you should do is familiarize yourself with the layout of your ship. Determining whether or not there is a doctor onboard and where they are located is important.

Since germs and illnesses spread easily aboard cruise ships, you will need to protect yourself. This protection involves washing your hands on a regular basis. When most individuals are at home, they wash their hands after completing certain activities. In addition to eating and using the restrooms, you should wash your hands after using all public facilities. If you find it difficult to wash your hands regularly, you may want to obtain a package of disinfecting wipes. If used properly, disinfecting wipes can be just as effective as soap and water.

Cruise ships are popular for their relaxing atmosphere. This relaxation may include lounging. When lounging or taking a stroll, it is advised that you keep your shoes on at all times. As previously mentioned germs can easily spread aboard a cruise ship. Keeping your feet covered at all times will not only protect you from contagious illnesses, but it will prevent you from stepping on harmful objects.

Protecting Your Health While Aboard a Cruise Ship

When it comes to cruise ship safety, there are many individuals who are concerned with their physical safety. Physically staying safe aboard a cruise ship is important, but so is staying healthy. Due to a large number of individuals in a fairly small space, cruise ships have a number of health concerns.

The health problems and concerns found on cruise ships are often highly publicized. Despite this publicity, most cruise ships are fun, safe, and clean. Being concerned with your health aboard a cruise ship is one thing, but you should not let it prevent you from enjoying your vacation. Rather than letting the fear of getting sick ruin your vacation, you should take a few simple steps to ensure that your health will be the same as, if not better, when you first boarded.

Before your cruise ship is scheduled to set sail, you are urged to schedule a visit with your family physician. Be sure to inform him or her that you are planning a vacation aboard a cruise ship. Your physician will likely check your health to make sure you are fit for travel. In addition to examining your general health, it is possible that your physician may require a number of immunizations. These immunizations may offer you extra protection; therefore, you are urged to obtain them.

If you are taking multiple medications at the time of your cruise, you will want to make sure that you have an adequate supply. If you are running low on your medication, you may want to obtain a refill before you leave for your cruise. It is also likely that your doctor may request that you bring along extra medication. This request is often made to individuals who must take their medications to survive. In the event that you misplace your medication, you will have an extra supply to fall back on.

Taking health precautions before you board the cruise ship is important. However, it is also important to take precautionary measures after you board. These measures are simple; however, they may be vital to your health. One of the first things that you should do is familiarize yourself with the layout of your ship. Determining whether or not there is a doctor onboard and where they are located is important.

Since germs and illnesses spread easily aboard cruise ships, you will need to protect yourself. This protection involves washing your hands on a regular basis. When most individuals are at home, they wash their hands after completing certain activities. In addition to eating and using the restrooms, you should wash your hands after using all public facilities. If you find it difficult to wash your hands regularly, you may want to obtain a package of disinfecting wipes. If used properly, disinfecting wipes can be just as effective as soap and water.

Cruise ships are popular for their relaxing atmosphere. This relaxation may include lounging. When lounging or taking a stroll, it is advised that you keep your shoes on at all times. As previously mentioned germs can easily spread aboard a cruise ship. Keeping your feet covered at all times will not only protect you from contagious illnesses, but it will prevent you from stepping on harmful objects.

In addition to cruising along water, most cruise ships dock in a port. It is likely that you will be visiting a foreign country. In the event that you end up on foreign land, it is important that you carefully monitor what you eat and drink. If you notice drinking advisories, you are urged to follow them. Instead of drinking tap water, you may want to opt for bottled water or another bottled beverage. Many individuals are concerned only with the drinking water, but you should also be concerned with ice cubes and water in uncooked foods.

As you can see, the above mentioned precautionary measures are simple and easy to follow. Why put yourself at risk for illness; especially when it requires little or no effort to keep your health in good standing.

Safety on a Cruise Ship

A cruise is one of the most enjoyable ways to see the world, whether you are cruising in the Caribbean or the Mediterranean! After all, how else can you see multiple cities and countries and unpack and repack only once instead of moving to a different hotel each night? You want to take a cruise, but with all of the news reports about things happening on cruise ships, how safe can it be? In reality, cruise ships have had an overall excellent safety record, and considering the number of people they move each year, cruise ships have a very low rate of crime. By using common sense and logic, your trip can be quite enjoyable.

Make sure that you do not pack valuables in your luggage, as you will check your bags at the pier when you arrive, and they will not be delivered to your room for at least a few hours. Most, if not all, luggage goes thrrough x-ray machines, so if you have left a laptop or jewelry in your suitcase, it will show up on the x-ray if someone is looking for a "golden opportunity," Most importantly, do not pack medications in your checked bags. This will also apply when your cruise ends, as you leave your luggage outside your cabin the night before you dock. Use the safe provided in your cabin to keep your valuables and tickets secure. If you aren't comfortable with the safe in your cabin, use the ship's safe located at the pursers office.

Security on all ships is very strict – each passenger is issued an ID with his or her name and cabin number printed on the card and photo imbedded into the data strip. Every time you embark or disembark, you go through security where staff checks to make sure the photo belongs to the person boarding or leaving.

One benefit of being on a cruise is that you do not have to get behind the wheel of a car after an evening out, but one's judgment, perception and behaviour are still compromised by excessive alcohol consumption. Staying aware of your surroundings and making sure that you do not wander around the ship in the middle of the night by yourself will generally help keep you out of trouble.

When in port, pay attention to your surroundings. Be aware that walking through an unfamilliar city looking confused or lost can be like a beacon to pickpockets and thieves. If you do not want to explore on your own, taking a tour offered through the cruise line will give you the chance to see the sights AND meet your fellow passengers.

The key to a successful cruise is to relax and enjoy the sights. By taking a few precautions, you will be all set to stand at the railing and wave to spectators on the pier as you sail off into the sunset.

Safety on Celebrity Cruise

Throughout its history, Celebrity Cruises® has provided its guests with a wide variety of cruise experiences that lead to exceptional vacations. Whether it's our professional and friendly crew, our unmatched fleet of ships, our almost endless variety of activities, our extensive list of beautiful and exotic destinations and excursions, our exciting entertainment, or our wide variety of delicious food, you will find many reasons why your Celebrity Cruises vacation will be one of the best you have ever experienced.

You may also find that one of the most amazing parts of your cruise experience will be the many rich and varied cultures represented among our guests. This diversity offers a chance to learn about different parts of the world and to make new friends. But just as a new friend can enhance your vacation experience, some guests may behave in ways you find unusual or undesirable. Since our guests do come from different and diverse backgrounds, it is important that everyone have a common understanding of the behavioral standards in place on Celebrity Cruises ships.

This Guest Conduct Policy is intended to help ensure that all guests are able to participate in a safe and enjoyable cruise experience and sets forth standards of conduct for guests to follow throughout their Celebrity Cruises vacation, including transfers to and from ships, inside terminals, while onboard, at ports-of-call, during shore excursions and at our private destinations. This policy is not intended to be all inclusive, and it is likely there will be conduct issues that it does not specifically address. In that event, as in all others, guests are expected to follow the direction of the ship's Captain who will take appropriate action to ensure the safety, security and well-being of our guests. In addition to this policy, guests are expected to comply with all applicable laws of the various countries that their cruise visits. Updates to the Guest Conduct Policy may be made between publications of this directory. Any updates to this Guest Conduct Policy are available for review on the Celebrity Cruises internet site

Safety and Security

Safety and security are everyone's responsibility. Should you become aware of someone being injured, or of unsafe or possibly illegal behaviour during your cruise vacation, it is very important that you immediately report this to the ship's management. This may be done by calling the ship's emergency telephone number listed in the Directory of Services in your stateroom; or by calling or going to the Guest Relations Desk. If you do not immediately report an injury or unsafe/illegal behaviour this delay may cause ship's personnel to be unable to effectively respond to the situation.

A reporting delay may also cause the ship to be unable to properly preserve information or evidence and may cause a delay in notifying the proper law enforcement or other Government officials. You can find additional helpful information in the "Security Guide" available on our website and onboard at Guest Relations.

Guest Conduct

Guest and Crew Interaction

Our crew members are friendly, outgoing and helpful, and they will do their very best to make your vacation as enjoyable as possible. Please do not misinterpret their friendliness. Crew members are prohibited from engaging in physical relationships with guests.

Crew members are not permitted to socialize with guests beyond their professional duties, and are not permitted to be in guest staterooms, except for the performance of their shipboard duties. Guests are expected to respect these policies and are similarly prohibited from engaging in physical relationships with crew members. Guests are not permitted in any restricted or crew area of the ship, including crew staterooms and corridors.

Verbally Abusive or Offensive Language

Verbally abusive or offensive language directed toward anyone, to include guests, crew members, governmental officials, or others is not permitted.

Inappropriate or Abusive Behaviour

Inappropriate or abusive behaviour including uninvited physical contact, solicitation, harassment, vandalism, theft, violence, use of fake/false identification, underage drinking, providing alcohol to those under age 21, possession of illegal substances/ items or any other illegal or offensive conduct is not permitted.

Unsafe Behaviour

Sitting, standing, laying or climbing on, over or across any exterior or interior railings or other protective barriers, or tampering with ship's equipment, facilities or systems designed for guest safety is not permitted. Guests may not enter or access any area that is restricted and for the use of crew members. Any other unsafe behaviour, including failure to follow security instructions, is not permitted.

Discourteous or Disruptive Behaviour

Pool, deck and theater chairs may not be reserved. Topless sunbathing is not permitted except on selected itineraries, and then it will only be allowed in designated areas. (Contact Guest Relations for Dress Code details.) Boom boxes or loud radios are not permitted. Roller blades, roller skates, skateboards, scooters, surfboards, bicycles, and similar items may not be utilized onboard; except for mobility aids related to special needs, as approved by the ship's medical staff.

Smoking

For the comfort and enjoyment of our guests, our ships are designated as non-smoking; however, we recognize that some of our guests smoke. Therefore, to provide an onboard environment that also satisfies smokers, we have designated certain areas of the ship as "smoking areas." Generally, smoking is not permitted except in designated areas. (Contact Guest Relations for additional details). This policy includes smoking-like products such as electronic cigarettes. Smoking is not permitted inside any guest stateroom; nor on any guest veranda. If a guest is in violation of this policy, a cleaning fee of $250 USD will be applied to their SeaPass account and the guest may be subject to further action pursuant to the "Consequences Section" of this Guest Conduct Policy.

Celebrity Cruises kindly asks all guests to please observe the non-smoking areas and to refrain from smoking cigarettes, pipes or cigars in any of the non-designated areas. These requests are made to provide a comfortable shipboard environment for everyone. Where permitted, cigarettes, cigars and pipe tobacco must always be properly disposed of and never thrown overboard. A guest must be at least 18 years of age to purchase, possess or use tobacco.

Curfews

Ship's management reserves the right to enact curfews on an individual, group, or ship-wide basis, if in the sole judgment of the ship's Captain, such steps become necessary to ensure guest or crew safety.

Disembarking from a Ship

Parents or guardians must not permit any guest in their care under age 18 from leaving a ship in any port without responsible adult supervision.

Parental and Guardian Responsibility

For purposes of this Guest Conduct Policy, a minor is defined as anyone under the age of 18. A young adult is defined as anyone ages 18, 19 or 20. Parents and guardians are responsible for the behaviour and appropriate supervision of their accompanying minor(s) and young adult(s) throughout their vacation. This obligation applies during transfers to and from ships, inside terminals, while onboard, at our ports-of-call, during shore excursions and at our private destinations. This responsibility applies at all times, regardless of whether the parents and guardians are physically in the company of their minor(s) and young adult(s).

Alcohol

Celebrity Cruises guests are expected to be responsible for their actions at all times, including during transfers to and from ships, inside terminals, while onboard, at our ports-of-call, during shore excursions and at our private destinations. Consuming alcohol to excess impairs one's judgment and reduces one's ability to recognize and avoid potentially dangerous situations. Guests who choose to consume alcohol must do so responsibly. The ship's staff may refuse to serve alcoholic beverages to any guest who does not consume alcohol responsibly. Ship's personnel may request verification of a guest's age to verify they are of age to consume alcohol pursuant to this policy. Any guest that violates this alcohol policy, will be considered for disciplinary action pursuant to the "Consequences Section" of this Guest Conduct Policy, and may lose their privileges to use the disco or other areas or facilities of the ship.

On Celebrity Cruises, with the exception of embarkation day when guests are permitted to bring onboard with them two (2) bottles of wine (which are subject to a corkage fee) per stateroom;

guests are not permitted to bring alcoholic beverages onboard. Alcoholic beverages that are purchased from onboard shops or in ports-of-call (which must be presented to security upon re-boarding), will be secured by ship's personnel and delivered to the guest's stateroom just prior to the conclusion of their cruise vacation. Security may inspect containers (water bottles, soda bottles, mouthwash, luggage, etc.) at any time. Guests who are under the permitted drinking age will not have alcohol returned to them.

The minimum drinking age for all alcoholic beverages on Celebrity Cruises ships is 21. In certain circumstances, where local laws permit or require, Celebrity Cruises may modify this policy to permit the minimum drinking age on a ship to be less than age 21, but never less than age 18. Pursuant to such a modification, Celebrity Cruises may also require parent/guardian authorization.

A decision to modify policy and permit guests below the age of 21 to consume alcohol onboard during a sailing will be made by Celebrity Cruises at the corporate level (and not by an individual ship). Such a decision will be communicated to the ship prior to guest boarding. Guests may contact Guest Relations or refer to Celebrity Today for specific minimum drinking age information on their cruise vacation.

For purposes of complying with the minimum drinking age requirements, a guest's age is established upon boarding at the beginning of the cruise vacation. If a guest celebrates their birthday during the cruise vacation, and thereby becomes of age to consume alcohol, the guest may thereafter ask the Guest Relations Manager to modify ship's records to permit their consumption of alcohol during the remainder of the vacation. The guest will be required to appear at Guest Relations to present a government issued form of identification to permit verification of their age.

In certain situations, Celebrity Cruises may permit a parent or guardian who is sailing with his or her 18 to 20 year old young adult son or daughter, (or other young adult under their legal supervision) to sign a waiver allowing the young adult to consume alcoholic beverages onboard. In such situations, the authorizing parent or guardian and the young adult must agree to be responsible for ensuring the young adult will consume alcohol responsibly and otherwise comply with the Guest Conduct Policy, including among other things, not providing alcoholic beverages to any other person, regardless of age.

On cruise vacations boarding in a country where the legal drinking age is lower than 21 and where a young adult (age 18, 19 or 20) is not traveling with a parent or legal guardian, they will generally not be permitted to consume alcohol. There may be exceptions made to this restriction in areas of the world where local laws require and Celebrity Cruises concurs. Details on such exceptions can be obtained from Guest Relations.

No guest under age 18 may possess or consume alcohol at any time, while onboard or at our private destinations. No guest under age 21 may possess or consume alcohol at our private destinations. Any guest who goes ashore and consumes alcohol (whether under the supervision of a parent/guardian or not, is responsible for ensuring they consume responsibly and retain their ability to recognize and avoid potentially dangerous situations when they return to the ship. Parents/guardians are reminded they are responsible for the actions of their child/young adult at all times while on a Celebrity Cruises cruise vacation.

Guests who violate any alcohol policy, including but not limited to underage drinking; providing alcohol to minors or young adults; possessing, concealing or attempting to conceal alcoholic items in their luggage, when boarding, or while onboard; engaging in alcohol drinking games; or failing to consume alcohol responsibly, will be considered for discipline under the provisions of this policy.

Prohibited Items

Items with Heating Elements or Open Flames

Certain items that generate heat or produce an open flame are not permitted on ship. This includes clothing irons, hotplates, candles, incense and any other item that may create a fire hazard. Curling irons and hair dryers are allowed.

Drugs or other Illegal Substances

No illegal drugs or other illegal substances are allowed onboard or may be utilized during a Celebrity Cruises vacation, including during transfers to and from ships, inside terminals, while onboard, at our ports-of-call, during shore excursions or at our private destinations. Illegal drugs or substances will be confiscated and appropriate action taken, which may include removal from the ship and involvement of appropriate authorities. In addition, foreign governments at Celebrity Cruises ports-of-call have strict laws that address drug possession. Guests found in violation of such laws are subject to arrest and prosecution by the foreign jurisdiction (and perhaps United States or other authorities as well) and may be prevented from re-boarding the ship.

Weapons, Explosives or other Dangerous Items

No weapon, explosive, or other item that presents a risk of harm to persons or property, are permitted onboard. They will be taken by ship's Security and appropriate action taken. Items such as dive knives must be reported to ship's personnel at embarkation. They may be permitted onboard, but held in safe custody by the ship's Security staff when not in use ashore.

Health and Environment

Hand Washing

Guests are strongly encouraged to wash their hands with soap and hot water after using the restroom and before eating or handling food. Medical experts say this is one of the best ways to prevent illnesses from starting or spreading..

Illnesses and Isolation of Guests

Celebrity Cruises follows practices and maintains policies that seek to prevent illnesses from affecting our guests. One of the best ways to prevent the spread of contagious illnesses like gastrointestinal viruses, colds and flu is to wash your hands thoroughly for at least 20 seconds with soap and hot water after using the restroom and again before eating anything. In the event you experience symptoms that indicate a gastrointestinal illness, such as diarrhea or vomiting, or know of someone experiencing these symptoms, you must advise the medical staff immediately. In addition, some gastrointestinal illnesses remain contagious up to 72 hours or more after symptoms subside; therefore, if you or someone you know experienced such an illness just prior to your cruise, you must immediately notify the ship's medical staff. This will permit the crew to take steps to reduce the chance that your illness will spread to others on the ship. Failure to immediately report a contagious illness to the ship's medical staff, or to accurately describe its onset, greatly increases the likelihood the illness will spread to others and is a violation of this Guest Conduct Policy. In the event of a contagious illness, the ship's crew will take steps to curtail its spread to other guests, including if necessary, steps set forth in the Consequences Section below.

Fitness to Travel Safely

Guests with disabilities are not required to travel with another person as a condition to traveling on our ships. However, all guests must consider that crew members are not required to assist guests with personal tasks or personal hygiene needs. (E.g. assisting with eating, dressing, toileting.)

Therefore guests requiring assistance with these functions should consider these needs when planning a cruise vacation. If there is a question as to a prospective guest's ability to independently provide for their personal needs or hygiene without being a danger to themselves or others, and the guest plans to travel alone, this situation must be discussed with the company's Access Department prior to booking a cruise vacation. The Access Department can be reached at Special_Needs@Celebrity.com or by calling 1.866.592.7225. This will allow an individualized assessment of the guest's fitness to travel for the duration of the cruise without personal assistance. Unaccompanied guests attempting to board or found onboard without the ability to independently care for their personal needs will be evaluated to determine if they are fit to safely travel without assistance. Guests who are found unfit to travel may be denied boarding or removed from the ship at the next port-of-call.

Medication

Guests must ensure that they bring an adequate supply of any medication they require for the entire duration of the cruise. The ship's medical centres may not have a supply of any needed medication and not all medication will be available in the ports-of-call. Guests must not pack medication in checked luggage to ensure you always have immediate access if needed.

Environment

Age Policies

Facility	*Age Requirement*
Thalassotherapy Pool*	Age 16 and over*
Fun Factory Facilities	Ages 3 through 17, with age-specific sections.
Theatre	Under age 16 must be accompanied by a parent or adult guardian
Pools**	Guests must be fully toilet trained to enter the pools or whirlpools
Whirlpools **	Under age 16 must be accompanied by a parent or adult
Ports of call	Under age 18 must be accompanied by a parent or adult
Bingo	Under age 18 must be accompanied by a parent or adult
Fitness Centre	Age 16 and over
Aqua Spa	Age 18 and over
	Ages 13 through 17 may only be present while participating in scheduled treatments
Adult Night Club/Disco	Age 18 and over
Casino***	Age 18 and over (except Alaska***)
Specialty Restaurants	Contact Guest Relations Desk for age restrictions.

* During inclement weather and select hours, the ship may permit the Solarium Pool (if covered) to be used by persons under age sixteen (16) provided they are directly supervised by a parent or guardian.

** United States Public Health Department Policy (USPH) prohibits persons who are not toilet trained or use diapers, pull-ups or swimmers from using pools or whirlpools.

*** For all Alaska itineraries, guests must be twenty-one (21) years of age or older to gamble in the Casino.

Celebrity Cruises' Save the Waves Program and Trash/Waste Disposal

Celebrity Cruises maintains a Save the Waves Program that focuses on three key principles: Reducing the creation or generation of waste materials; recycling as much as possible; and ensuring proper disposal of remaining waste. Trash should be properly disposed of in containers provided throughout the ship or in wastebaskets provided in each stateroom. Trash or other foreign objects should never be flushed down a toilet and guests are not permitted to discard any item overboard.

Consequences

Failure to act in accordance with this policy may result in:

- Intervention by Security, other management personnel, or law enforcement
- Removal of certain onboard privileges, which may include being detained, quarantined or confined in a cabin or holding cell
- Confiscation of illegal/contraband/prohibited items (may be turned over to law enforcement authorities)
- Denial of boarding on the current or any future Celebrity Cruises cruise
- Reporting of incidents to government and law enforcement authorities for follow-on legal action
- Removal from a ship at the next port of call. Guests removed from a Celebrity Cruises ship pursuant to this policy, are responsible for their own accommodations and transportation home, at their expense. Documentation requirements for re-entry into the guest's home country are also the responsibility of the guest.

5

Security and Social Issue

How Working on a Cruise Ship is Emotionally as Well as Physically Demanding

Thirty years ago, a light-hearted American TV programme, "The Love Boat", painted a picture of sun, fun and romance on board cruise ships. Since then, cruise ships have become bigger and grander, carrying up to 3,000 passengers, mostly from North America and Western Europe, in huge floating hotels sailing the world.

For the past 20 years, these well-off tourists (Cruises with few exception is for mass tourism, it is the cheapest way for holiday of all) have made cruise shipping the fastest growing sector of the world maritime industry, with an average growth rate of 9.6 per cent. There are now about 250 cruise ships over 1,000GT sailing the oceans.

But the passengers are kept in blissful ignorance about the people who cater for their every need. Often paid derisory wages of around US$50 a month – less than the price of a pair of sunglasses in the onboard shop (It is not true at all 50 Usd is the minimum paid cash, while the company guaranteed wages which are well above that figure, no one on board get less than 800 dollars as minimum – they work 12-hour days, seven days a week. Instead of improving, work for seafarers on cruise lines has become more demanding as companies decrease their ratios of crew to passengers (yes in some cases, while on others companies compete against eachothers offering higher pax/crew ratio. And to eke a living, they have to keep smiling.

Only by keeping a smile on their faces and making sure they please the passengers are they likely to receive the tips that make up their wages. This regime is used by management to make sure that seafarers give every ounce not just of their physical energy but their emotional energy to please the company's customers. Recruitment decisions are made on stereotypes, based on which races and nationalities are considered the most friendly, hospitable and cheerful.

But these "below stairs" hotel and catering workers will never smile enough to attain prestigious roles with more status and money. They remain the preserve of mostly white, Western Europeans, to sustain the nostalgic image of trans-atlantic liners. (The biggest cruise companies have mixed race hotel crew eastern European are a minority compared to Asian and Central American.)

The commercialisation of emotional labour by cruise lines is the subject of a report from the ITF-funded Seafarers' International Research Centre (SIRC) at Cardiff University, UK. The centre's Deputy Director, Dr Minghua Zhao, has conducted research into the working and living conditions of women seafarers on cruise ships. Emotional labour was not part of the original brief – but, as Dr Zhao discovered, the ability to maintain a happy, smiling face and attitude, despite bad pay and exhaustion, is now demanded by cruise lines.

"Smile is a vital component of the product and service cruise lines promise to deliver to meet their customers' demand for leisure. Seafarers' smiling faces, men and women of various colours, are displayed on travel agents' shelves and permeate cruise lines' brochures, advertisements in newspapers, travel shop series on television and so on.

"These smiling faces are presented to convey a message and to leave an impression: on cruise ships, seafarers enjoy their work as much as passengers enjoy their fun and leisure. 'It's a heaven not in this world!' as one of the major cruise lines says in its television advertisement.

"In fact, shipboard labour is extremely hard for seafarers on cruise ships. This is probably the only point agreed by both trade unions and ship owners. Either in trade unions' publication or in crewing agencies' pamphlets or in shipping companies' brochures, seafaring labour in hotels and catering departments is described as 'involving work for 12 hours a day and seven days a week', and 'the work is conducted in a confined space, far away from your family and friends for most part of the year'. The only party deliberately kept away from this knowledge is the customers."

Cruise lines' recruitment managers admit that they favour certain races and nationalities for their apparent ability to keep smiling. One told Dr Zhao: "Interviews... give us an opportunity to meet the applicants and evaluate their personality and ability to work with people. Smiling is certainly an extremely important area we will look into. After all, this is a hospitality industry and smiling faces are more welcomed by our guests. We use many Indians on our vessels. Believe me, they work hard and they always smile."

Another said: "Somehow, the Asian seafarers, especially the Filipinos, can smile very nicely. They seem to have born with a wonderful service culture. They always greet the guests and always smile. And they do it so naturally. We get very good feedback from our guests about these seafarers. They always look energetic, positive and cheerful, even after nine months at sea! In comparison, European seafarers seem easy to get fatigued. After four months, their fatigue will show, and our guests don't like to see those who serve them look tired and unable to smile."

The stereotyping also works in reverse. Eastern European people are believed by shipping managers to be dour. Dr Zhao said: "Managers tend to make particular notes about seafarers from Eastern European countries, accusing them of 'not being able to smile' or 'they are always so rigid'. Many of them admit that they would give priority to employing Asian seafarers, as long as seafarers from both regions are available in the labour market and their rates remain the same low." Only when profits are affected – such as by the high cost of air fares to bring Asian seafarers to ships docked in Europe – are cruise lines prepared to overlook their own prejudices.

But no matter how brilliant or enduring their smiles, or how good they are at their jobs, certain races and nationalities will certainly be overlooked for promotion or jobs at 'key points', with most contact with the passengers. The report says: "While the managers are predominately white male, Asian seafarers experienced in hospitality industry are increasingly employed as waiting seafarers. Those from Thailand, especially men, are particularly popular with ship managers because, in addition to their experiences, they are assumed to be able to smile 'naturally' and to have 'born' with a right 'service culture', hence they are 'popular with passengers', as we were told by ship managers.

"However, we found, through our shipboard observation of the operation process in bars and restaurants and our in-depth interviews with seafarers serving in these areas, that these Asian seafarers have to work even harder on their emotion in order to deliver the 'proper' service which makes them popular with passengers. Sam, the Thai headwaiter, for example, was one of the few seafarers who were most popular with the passengers and collected most tips at the end of almost every cruise. His main station was the main restaurant, where he worked for 11 to 12 hours a day and seven days a week, as every other waiter or waitress does. What distinguished Sam, however, was his 'brilliant memory' and his 'broad smile'. He was always at the restaurant entrance, greeting every passenger with a bright 'good morning' or 'good afternoon' and a broad smile. He remembered many passengers' names, noticed their likes and dislikes, asked about their families, and sometimes flirted a bit with women passengers.

"But this apparently forever-happy-energetic-and-smiling headwaiter is transformed into another person with the progress of our two-hour interview in a quiet office under the bridge. He becomes serious and shares with us how he feels about the job:

It's a very hard job. I always feel exhausted when I finish it and come down to my cabin. You may not know the nature of hospitality industry. The restaurant is a stage, a show. You are an actor. Believe me, you can be totally drained just by greeting people, chatting with them, smiling to them, and things like that. As an Asian, I have to work harder, I have to make more efforts to please the passengers. Sometimes, I feel I have to flirt a bit with the ladies, because I find it works. People are generous if they like you. "Sam and another Thai headwaiter are the two most popular headwaiters in the main restaurant, the best tip-earners among all the seafarers, and have been selected many times as the Best Seafarer of the Month in their department. However, when the post of restaurant manager became available, it was another European headwaiter who was promoted to the position."

Dr Zhao concludes: To seafarers on cruise ships, the labour is as hard as before. Indeed, it has become harder. Since the 1990s, ship owners have begun to compete with each other to build larger and larger ships. The 'economy of scale effect', resulting in increased profit, is certainly behind the ship owners' and operators' strong interest in large cruise ships. Both capital cost and operating cost per passenger decreases rapidly with the increase of ship size. With bigger ships, crew costs only increase marginally.

During the same period, the crew-passenger ratio has kept growing with the increase of the ship size. Traditionally, the typical ratio was 1:2 to 1:2.5 depending upon the position of the ship in the cruise market. Each cabin seafarer looked after 10 to 12 cabins, as observed by Maxton-Graham 20 years ago. Now, it has become 1:3 on most ships and can be 1:4 or even higher. It means that the seafarer has to clean more cabins, remember more names and smile to more passengers. They have to labour harder for longer hours.

But, changes in this aspect did not happen overnight. The intensification of seafaring labour has happened gradually, and subtly in many cases. On board today's ship, the single most striking difference from the early 1980s is that chances are getting smaller of seeing the same work conducted by Norwegian or seafarers from other developed countries. Now, these tasks are conducted by seafarers from developing countries, from Asia, from Eastern Europe, from Latin America and so on. Placed as front line workers at the bottom of the ship hierarchy, these seafarers are bearing the brunt of the restructuring of the world shipping.

Safety

Staying safe while you are at sea is as important as staying healthy.

Working at sea can expose you to particular risks. Seafarers have a one in 11 chance of being injured on their tour of duty – much higher than other occupations.

How do I Stay Safe on Board?

Slips, trips and falls are the biggest hazard at sea – with the danger of fractured or sprained limbs. Decks and gangways are the most injury-prone areas so be sure to watch your footing at sea. It is also important to exercise, eat healthily and avoid fatigue. The fitter you are, the more alert you will be.

Are there Guidelines Available?

If you work on cruise ships or passenger vehicles, you also need to take the safety of passengers into account. ITF has specific policy advice for best practice in ensuring the safety of yourself, your crewmates and your passengers.

The ILO code of practice for accident prevention on board ship at sea and in port provides practical recommendations and useful guidance. These include the establishment of a safety and health committee on all ships, with elected seafarers' safety representatives. Further information is available on the right of the page.

ITF has produced a training video for ships' safety representatives. This aims to assist crew and safety reps to improve on-board safety culture, and how to avoid and deal with any accidents. The video is available for viewing as a link on the right of this page.

ITF is developing a specific policy on health and safety in shipping. This will provide practical guidance to ITF seafarers' affiliates, their shop stewards, shipboard representatives and members on safety and health in shipboard work.

Why do Accidents Happen at Sea?

Accidents can happen when ships are under pressures to meet unrealistic schedules with too few crew. Owners sometimes cut corners by not maintaining the ship and its equipment, increasing the risks. Ships registered under Flags of Convenience have fewer regulations, and greater potential for unsafe practices.

The risks can be minimised where crews are fully trained in the safe running of the ship. Ship owners should also provide medical care and social support.

Have you Had an Accident?

Seek advice on any first aid issue – radio medical advice is available throughout the world, in port or at sea, and your ship's master is obliged to be fully conversant with radio medical procedures.

Make sure you report any accident on board. This will help maintain the safety of the ship for current and future crew members. You will also stand a better chance of receiving any compensation.

The ITF has dedicated claims handlers to assist seafarers and their families in claims for compensation following injury or death. If you have suffered an accident, please contact the ITF and/or your union without delay, using the link on the right of this page.

We will advise you of the best way in which your claim can be handled and, if necessary, will take it up directly with the agents and owners to negotiate the best possible compensation for you.

U.S. Coast Guard Cruise Ship Consumer Fact Sheet

U.S. Coast Guard

Vessel Safety. Ocean-going cruise ships of U.S. registry must meet a comprehensive set of Coast Guard safety regulations and be inspected annually by the Coast Guard to check for compliance. The safety regulations cover such things as hull structure, watertight integrity, structural requirements to minimize fire hazards, equipment requirements for lifesaving, firefighting, and vessel control, and requirements pertaining to the safe navigation of the ship. If the ship passes its annual inspection, it is issued a Coast Guard Certificate of Inspection valid for one year. The certificate must be displayed where passengers can see it.

Today, nearly all the ocean cruise lines employ passenger ships registered under flags of various foreign countries. (Note: The law requires that cruise-ship advertising in the U.S. disclose the country of registry.) Each ship is subject to the vessel inspection laws of the country in which it is registered. However, as a condition of permitting the vessels to take on passengers at U.S. ports, the U.S. Coast Guard requires the ships to meet the International Convention for the Safety of Life at Sea (referred to as SOLAS.)

SOLAS and other international regulations also require compliance with stringent regulations regarding structural fire protection, firefighting and lifesaving equipment, watercraft integrity and stability, vessel control, navigation safety, crewing and crew competency, safety management and environmental protection.

To insure compliance with SOLAS, the Coast Guard examines the ship when it first goes into service at a U.S. port, with quarterly checks thereafter. The examinations emphasize structural fire safety and proper lifesaving equipment. Fire and abandon ship drills conducted by the ship's crew are witnessed, and operational tests are made on key equipment such as steering systems, fire pumps, and lifeboats. The Coast Guard has the authority to require correction of any deficiencies before allowing the ship to take on passengers at the U.S. port. The records of these examinations (called Control Verification Examinations) are open to the public at the Coast Guard Marine Safety Office (MSO) which conducted the examination.

Crewmember Competency. On U.S. passenger vessels, licensed individuals and crew must meet standards for experience and training set forth in Coast Guard regulations. The Coast Guard can revoke or suspend the individual's license or merchant mariner's document for acts of misconduct or incompetence. On foreign-flag cruise ships trading in the U.S., SOLAS requires the vessel to be sufficiently and efficiently manned. The officers' licenses and the vessel's compliance with manning standards are checked as part of the Control Verification Examination.

Medical Care and Services Are Not Covered Coast Guard regulations and SOLAS generally focus on requirements for the safe navigation and design of the cruise ship itself. There are many aspects of the daily care of passengers and their on-board accommodation that are not covered by Government regulation. One example of this is medical care. The Coast Guard does not require that passenger vessels carry a ship's doctor. Most, if not all ocean-going passenger vessels today do provide a doctor and medical facilities in order to offer attractive and competitive service. Passengers should realize that the quality of their medical care is not guaranteed by Coast Guard regulations. If you are concerned about this aspect of life aboard a cruise ship, contact the cruise line or travel agent for the particulars of medical services provided, both at sea and while visiting foreign ports.

Emergency Drills. Coast Guard regulations and SOLAS require that the master of an ocean cruise ship periodically hold fire and lifeboat drills. They are intended not only to give the crew practice, but also to show the passengers how to act in the event of an emergency at sea. Passengers should participate fully in these drills. The timing and frequency of the drills depends in large part on the length of the voyage. On voyages that will last more than one week, the first drill will be held before the ship gets underway (passengers who embark at the last minute sometimes miss this drill), with additional drills at least once a week thereafter. On voyages of one week or less, the drills must be held within 24 hours after leaving port.

Coast Guard and international regulations also require a notice to be posted conspicuously in each passenger cabin or stateroom. The notice explains the following: How to recognize the ship's emergency signals (alarm bells and whistle signals are normally supplemented by announcements made over the ship's public address system); the location of life preservers provided for passengers in that stateroom (special life preservers for children will be provided, if necessary, by the room steward); instructions and pictures explaining how to put on the life preserver; and the lifeboat to which passengers in that stateroom are assigned. (Note: Passengers need not be alarmed if they discover that the total number of person's on board a cruise ship (passengers + crew) exceeds the total capacity of the ship's lifeboats. Modern cruise ships carry a variety of survival craft. Passengers are invariably assigned to lifeboats or similar survival craft. The total capacity of all the survival craft on board will exceed the total number of persons on the vessel).

When fire and lifeboat drills are held, crew members from the stewards department are generally responsible for assisting and directing passengers in the drill. Direction signs showing the path to reach lifeboats are posted in passageways and stairways throughout the ship. The crewmember in charge of each lifeboat will muster the passengers assigned to that lifeboat, and give passengers any final instructions necessary in the proper method of donning and adjusting their life preservers. If there is any portion of the emergency procedures the passenger doesn't understand, they should question the crew until the instructions are clear and completely understood.

Sanitation and Cleanliness. Oversight of sanitary conditions on passenger vessels is the responsibility of the U.S. Public Health Service (USPHS). The USPHS conducts both scheduled and surprise inspections of passenger vessels in U.S. ports. The inspections focus on proper sanitation for drinking water, food storage, food preparation and handling, and general cleanliness. The USPHS will provide the public with results of inspections on individual vessels, and take reports of unsanitary conditions on individual vessels.

Vessels Are Responsible for Proper Trash Disposal. As part of its mission to protect the marine environment, the Coast Guard enforces regulations regarding ocean dumping from vessels. The regulations make it illegal to dump plastic refuse and garbage mixed with plastic into any waters, and restrict dumping of non-plastic trash and other forms of garbage. These regulations apply to all U.S. vessels wherever they operate (except in waters under exclusive jurisdiction of a State), and to foreign vessels operating in U.S. waters out to and including the Exclusive Economic Zone (200 miles off shore.) Terminal Security. In accordance with Federal regulations, terminal operators and cruise lines share the primary responsibility for shoreside and shipboard security of passengers. The Coast Guard examines all security plans and can require improvements in their security measures. Passengers embarking on international voyages may expect to have their baggage searched or passed through screening devices before boarding. The terminal operator and cruise line have strict procedures for passenger identification and visitor control. Passengers who wish to have friends visit the ship prior to sailing should check with the cruise line well in advance. All these security measures are designed to prevent the introduction of unauthorized weapons and persons on the cruise ship.

Financial Responsibility. The Federal Maritime Commission requires that operators of passenger vessels carrying 50 or more passengers from a U.S. port must be financially capable of reimbursing their customers if the cruise is cancelled. The Commission also requires proof of ability to pay claims arising out of passenger injuries or death for which the ship operator may bear some liability. It is important to understand that the Commission does not have the legal authority to automatically secure these financial settlements for individual consumers. If a cruise is cancelled, or there is an injury incurred during the cruise, the consumer will have to initiate action on his or her own behalf against the cruise line.

Shipboard Workplace Code of Conduct

It is a comprehensive, industry-wide commitment to safety, security and fair treatment of crewmembers in the shipboard work environment.

Compliance with Regulation

The United Nations has established two specialized agencies to deal with international maritime issues: the International Maritime Organization (IMO) based in London and the International Labour Organization (ILO) based in Geneva. Both organizations have adopted conventions dealing with the safety, health and welfare of seamen. The most widely adopted convention that governs shipboard labour and crew employment practices is ILO 147. ILO 147 has been ratified by 42 countries, including the United States.

Normally the applicable law will be that of the flag state of the ship on which the employee is serving, but the law of the employee's country of residence or, in certain situations, the law of the port state in which a ship is berthed may also be relevant. Port states, such as the United States, have the authority to enforce crew safety regulations. The U.S. Coast Guard has the authority to ensure that the provisions of ILO 147 are adhered to on all ships arriving at United States ports. The United States Public Health Service inspects passenger vessels that call on U.S. ports to ensure the highest public health standards.

General Conditions of Employment

As a guiding principle, member cruise lines conduct their worldwide operations according to the highest standard of business integrity and fair treatment of employees. All personnel, shipboard and shore side, have the basic right to be respected and treated in a fair and just manner at all times by superiors and fellow employees. As a global industry, the cruise lines recognize the value that different cultures bring to the industry, and encourage workforce diversity, and provide a workplace free of discrimination based on gender, race, religion, age, disability, nationality, sexual orientation, social or ethnic origin. All crewmembers will be 18 years of age or older.

Hours of Work

Hours of work and hours of rest will comply with flag state regulations and/or applicable collective bargaining agreements.

Living Conditions

All crewmembers receive room and board at no cost and live in a clean, well-maintained living environment. The crew areas on each ship shall be inspected on a regular basis by the ship's master or his representative.

Health and Safety

The shipboard work environment will be monitored and governed through safety and quality management systems. Cruise line sponsored medical care is provided for all crewmembers working on board. Crewmembers who become injured or ill while working are entitled to receive sick leave.

The work environment operates in accordance with Safety of Life at Sea (SOLAS), the International Safety Management (ISM) Code, the Standards for Training Certification, and Watchkeeping (STCW), and respective flag state requirements.

Seafarer Identification

ILO Convention 108 requires that ratifying countries issue identification documents to seafarers. The U.S. State Department oversees the issuance of work visas to foreign nationals who sail with ships into U.S. ports. U.S. embassy personnel in the crewmember's country of origin conduct background checks. A cruise ship's work force is pre-screened, and checks are conducted on prospective employees.

Recruitment

Responsibility for the recruitment and employment of seafarers who serve on cruise vessels may be undertaken by third parties, such as ship managers or crew managers, or by the cruise line.

When personnel administration or the recruitment of seafarers is delegated to a third party, it is important to ensure that the activities of such agencies are monitored for compliance by the agency with its obligations under international law.

Repatriation

The precise circumstances in which the employer will be responsible for repatriation costs should be included in the employment contract, employment manual or terms and conditions issued to the seafarer.

Wages

Crewmembers are afforded wages that are competitive with comparable international pay scales and provided employment opportunities and compensation packages that are equal to or exceed similar positions in the nations from which crew are recruited. Depending on the position, there is also the opportunity for many crewmembers to earn gratuities. Additionally, policies include recruiting at the trainee level and providing education so crewmembers obtain work skills that facilitate promotion and career development.

Security

The cruise industry places the highest priority on the safety and security of its passengers and crew. Every person onboard a cruise ship, from the captain to the cleaning staff and all guests, will be placed on an official manifest.

Passengers and crew may embark or disembark only after passing through security. Once the ship is underway, access is strictly limited to documented employees and fare-paying passengers. Security personnel are employed onboard every vessel. Federal and state authorities have the right to investigate crimes onboard cruise ships. The FBI has the authority to investigate and prosecute alleged crimes in international waters involving

Americans. The U.S. Coast Guard has jurisdiction for inspection and enforcement of international safety and security standards for all ships calling at U.S. ports.

Complaint Resolutions

All ships recognize the right of employees to voice complaints. All ships will maintain an environment in which crewmembers may express complaints and obtain resolution of those grievances without fear of retaliation or retribution.

Crew Awareness

This Shipboard Workplace Code of Conduct will be made available to all crewmembers.

6

Policy and Governance Dimensions of Security

Terror by Sea: The Unique Challenges of Port Security

The United States always has defended its borders and natural resources from harm. Unlike threats of the past, however, maritime terrorism constitutes a multifront conflict both at home and abroad. It recognizes no borders and can originate from organized groups and individuals operating inside or outside the United States. The fight against maritime terrorism, unlike conventional threats, is a complicated struggle with neither an easy nor a rapid conclusion. The U.S. maritime jurisdiction represents a daunting challenge: more than 1,000 harbour channels with 25,000 miles of inland intracoastal and coastal waterways that serve 361 ports containing more than 3,700 passenger and cargo terminals. Annually, the U.S. Maritime Transportation System manages over 2 billion tons of freight, 3 billion tons of oil, more than 134 million ferry passengers, and an estimated 7 million cruise ship travellers. The dimensions of the extraordinarily difficult task confronting the United States centre on an estimated 7,500 foreign ships, manned by 200,000 sailors, entering U.S. ports every year. The more than 6 million intermodal cargo containers that enter annually represent two-thirds of the total value (as opposed to tonnage) of all U.S. maritime trade. (1) Monitoring the intricate waterways, ports, and interstate connections falls to the marine enforcement officers in the field.

Perhaps, a review of the ports on the Mississippi River can put the challenge these officers face into proper perspective. The lower 255 miles of the Mississippi River, from the mouth to Baton Rouge, contain 4 of the top 11 ports in the country. The Port of South Louisiana, located between New Orleans and Baton Rouge, encompasses a total of 54 miles, while the Port of New Orleans includes 33 miles of the river.

The Port of Greater Baton Rouge comprises 85 miles, and the Port of Plaquemines, located south of New Orleans, covers over 80 miles of the Mississippi River. Of particular interest is the Louisiana Offshore Oil Port, internationally known as the LOOP, located about 18 nautical miles south of Grand Isle. The nation's only offshore deepwater port, it can handle 100,000 barrels of oil an hour, or 2.4 million barrels per day, and services over 30 percent of this country's refineries. Such a valuable resource as the Mississippi River obviously requires protection from the threat of maritime terrorism.

Piracy Connection

A report on piracy and other criminal attacks at sea highlighted the vulnerability of shipping to terrorist attacks. "Today's pirate is tomorrow's terrorist and one can foresee that a nexus will ultimately develop between terrorist organizations and pirates. Both are acts of violence against innocent people, one done for monetary gains and the other perhaps for a political purpose. Since piracy is largely undertaken for a commercial gain, pirates will not hesitate to carry out a terrorist act if someone pays them enough and they have a reasonable chance of getting away." (2) This nexus cannot be ruled out. After all, if terrorist groups are to sustain their operations, they likely will tend toward mounting suicide attacks on commercial and military vessels for such purposes as—

* procuring alternative revenue for their main initiatives (i.e., simple piracy);
* carrying out suicide attacks on unsuspecting ships or ports in the spirit of September 11;
* smuggling weapons and explosives to their affiliates in other parts of the world, probably by hijacking a ship before renaming and otherwise disguising its identity and providing a new crew and manifest;

* seeking ransoms and trading hostages for members of their groups detained by authorities; and
* placing mines aboard innocent-looking tugs and barges, commonly found in the coastal and intracoastal waters and ports of the United States.

As an example, the Strait of Malacca, the world's busiest waterway, provides passage to more than one-third of the global trade. The number of vessels that ply the waters ranges from 50,000 to 330,000 each year. Geographically, the Strait of Malacca is 500 nautical miles in length and extremely narrow, especially along its southern half where it ranges from a mere 20 nautical miles wide to as little as 9 nautical miles at the southern end. A terrorist attack on a steamer could endanger the lives of its crew, threaten the safe passage of other vessels and the lives of their crews, and disrupt maritime traffic along the straits, crippling regional and international trade.

Post-September 11 World

Before the tragic events of September 11, marine enforcement officers received specialized training to combat drug smuggling and illegal immigration and to protect America's natural resources and environment. Today, however, because of the threat of maritime terrorism, the need for advanced, specialized, and tactical training of these officers has become far more critical.

As the tip of the spear, the U.S. Coast Guard and Customs and Border Protection (CBP) are tasked with the protection of America's ports. One of the Coast Guard's principal missions involves safeguarding U.S. ports and waterways, whereas the primary mission of CBP is to inspect cargoes and cargo containers entering U.S. ports. (3) To meet the unprecedented challenges facing them, the Coast Guard and the CBP have allocated additional workers to their respective port- and cargo-security missions and have initiated several programs designed to tighten security not only of port areas but also of merchant ships, the cargoes they carry, and their crews.

The United States has taken the common-sense approach of detecting and deterring potential threats long before they escalate into clear and present dangers. In the maritime arena, this requires "identifying and intercepting threats well before they reach U.S. shores." (4) This strategy supports national-level objectives while recognizing the uniqueness of the maritime environment, including the difficulty associated with the shared use of oceans and waterways.

Denying the use and exploitation of the maritime domain by terrorists as a means to attack U.S. territory, population, and critical infrastructure requires increased maritime domain awareness (MDA) and enhanced security operations. Currently, the U.S. maritime domain is protected via intelligence information and layered, multiagency security operations nationwide, including the strengthening of the security posture and reduction of vulnerability of American ports.

The first MDA consideration is the effective understanding of anything in the marine environment that could adversely affect America's security, safety, economy, or environment. Given the physical impossibility of patrolling the entire maritime domain, building a robust MDA capability can provide national leaders, operational commanders, and maritime stakeholders the information, intelligence, and knowledge needed to make operational and policy decisions. Enhancing maritime security operations includes the reality that resources are finite and security forces must be enduring, sustainable, and flexible enough to accommodate both local and regional requirements while remaining able to adjust to changing security levels. The shifting priorities have resulted in marine enforcement units evolving into special operations as opposed to merely patrol. For example, the Coast Guard created active-duty, multimission, mobile teams with specialized capabilities to close critical security gaps in the nation's strategic seaports. Representing just one element of the defence posture in U.S. ports, these teams augment existing Coast Guard units and law enforcement agencies as opposed to working alone. CBP also follows a forward-deployed strategy. The Container Security Initiative represents over two-thirds of all cargo containers shipped to the United States and requires, among other things, that incoming containers be screened before they depart for U.S. ports of entry, rather than after they arrive on American shores. Other national efforts include greater intelligence collection, new regulations, increased patrols, additional assets, and partnering with other law enforcement agencies and the maritime industry.

New Training

U.S. policies direct agencies in the maritime community to deter, detect, and defend against traditional and asymmetric attacks. Every federal, state, and local marine enforcement officer should have training that addresses the current trends in worldwide maritime terrorism. However, the agencies needing maritime enforcement training have limited qualified educational facilities that, in turn, have few vessels, specialized equipment, and vehicles with which to conduct training.

With this in mind, on April 6, 2004, the Federal Law Enforcement Training Centre, Driver and Marine Division convened a Curriculum Review Conference of the Advanced Marine Law Enforcement Training Program. Protecting over 90,000 miles of U.S. coastline has become an increasingly difficult task. Therefore, the conference proposed a new curriculum to address the contemporary tactical skills required in the marine enforcement specialty. In addition, the participants recommended changing the program name to the Antiterrorism/Port Security Boat Operations Training Program. Five days in length, the program trains marine law enforcement officers, security personnel, executive protection agents, supervisors of marine units, and others in the knowledge and skills needed to deal with contemporary threats in the maritime environment. The target audience for the revised program includes federal officers and agents with marine enforcement duties or supervisory responsibilities of marine enforcement units; state and local officers who support or work closely with federal agencies involved in marine enforcement; and U.S. Department of Defence personnel involved in harbour security, marine counterterrorism, and other marine tactical operations.

The new curriculum includes several innovative courses. The Strategic Maritime Threat provides marine enforcement officers with a comprehensive review of the history of maritime terrorism, crime, and piracy with a view toward identifying relevant, current trends. Upon completion, students will have a greater appreciation of the role of enforcement agencies to prevent, deter, and mitigate threats to security in the maritime environment. DHS Use-of-Force Policy familiarizes students with the use-of-force policy adopted by the U.S. Department of Homeland Security. It includes a detailed explanation of the use of force during vessel pursuits and port security evolutions.

Tactical Navigation Techniques teaches students to adapt navigational skills to a law enforcement situation. It concentrates on using radar in reduced visibility or at night to assist in pursuits and intercepts. It also trains students to use radar to assist them in enforcing security zones. Finally, the course introduces students to state-of-the-art integrated electronic navigation tools.

Defensive Boat Tactics shows students how to operate a patrol boat during maritime security operations. Knowing these skills will enhance students' abilities to work in cooperation with the Coast Guard during these critical evolutions, improve their confidence, and increase their performance of other marine enforcement duties.

Tactical Boat Pursuits introduces students to the tactics of single- and multiple-vessel pursuits, including positioning, screening, blocking, and handing off. Students practice these tactics during underway laboratories, while a pursuit scenario in a graded practical exercise evaluates their mastery of these skills.

High-Risk Boarding trains students how to conduct tactical boarding in high-risk situations. It covers tactical techniques, such as line of fire and triangulation, as well as assessing threats and reacting appropriately. It teaches students to deal with aggressive behaviour by methodically escalating the level of force necessary to bring a situation under control.

In the post-September 11 world, the threat of terrorist attacks reaches into every segment of American life, whether on land, sea, or in the air. Protecting this country's ports and waterways constitutes a daunting challenge due to the sheer magnitude of the task.

Recognizing the scope of such an undertaking, the Federal Law Enforcement Training Centre has created a new training program for marine law enforcement officers who must safeguard U.S. maritime interests. Providing these dedicated men and women with the best techniques and tactics for dealing with maritime terrorists will ensure that this country remains strong and committed to countering every threat, even if it is borne on the waves.

Dark Side of the Tourist Boom: Cruise Ship Controversies Cross Borders

Environmental and community activists in the Mexican Pacific resort of Zihuatanejo recently celebrated the cancellation of a cruise ship terminal that would have accommodated two thousand-foot "floating hotels" at once. Opponents feared a large pier on the town's main beach would contribute to bay pollution and ruin the quirky, small-town atmosphere that attracts return visitors year-after-year. For months, the People in Defence of the Bay coalition had staged protest rallies, posted on-line petitions and fired off letters to government officials – including planners at the federal Ministry of Communications and Transportation (SCT).

Set on Zihuatanejo's main beach, the multi-story Casa Marina was a hotbed of activism. Banners and signs opposing a new pier draped the building that houses locally-owned businesses as well as the Humane Society, where scraggly street cats and spooked pelicans are nursed back to health. Inside the edifice, merchant and green activist Natalia Rodriguez Krebs runs a store that offers eye-grabbing indigenous art, crafts and clothing.

Rodriguez and other locals have long received the cruise ships that anchor inside the bay one at a time, and shuttle passengers in small boats (known as tenders) to the existing pier where they run a gauntlet of day-tour operators. In the wake of the 9/11 attacks on the United States, Mexican marines guarding the municipal pier have joined the throng.

That is not the only influence that Washington's "war on terror" has exerted on this small Mexican town. Since international regulations drafted in the wake of 9/11 mandate that other boats maintain a fixed distance from the cruise ships, industry expansions imply an increasing privatization of public waterways. Hosting ever-bigger cruise ships also means transforming the natural environment and reallocating local resources. Bays are dredged, local water supplies tapped and garbage unloaded. At terminals, walkways are fenced off, access roads expanded for day-tour buses and local businesses sometimes threatened with relocation. Invasive species hitchhiking on the cruise ships can alter local ecosystems.

"To allow the construction of a marine terminal in the middle of Zihuatanejo Bay means the collapse of the town," declared environmentalist Silvestre Pacheco, project director for the SOS Bahia group.

The Business of Cruising

The dynamic that boiled over in Zihuatanejo is simmering in coastal communities across Mexico, the world's Numero Uno cruise destination. The number of ship passengers visiting Mexican waters doubled from about 3.2 million in 2000 to 6.4 million in 2007, according to the Florida Caribbean Cruise Association (FCCA), a trade industry group. Some 22 Mexican ports now host cruise ships, and federal officials plan to make the cruise business even bigger. Sailing to Mexico on vessels departing from California and Florida ports, U.S. and Canadian nationals comprise the vast majority of cruise ship tourists. In recent years, cruise vacations have become trendy. Even readers of the liberal Nation magazine and listeners to Air America are pitched cruises hosted by their favourite corporate-bashing celebrities. In fact, in a prematurely triumphant mood, the centre-left radio network plans a three-nation Obama victory cruise after the U.S. presidential election in November to Belize, Mexico and Guatemala.

Earmarking tens of millions of tax dollars for infrastructure development, the Mexican government justifies its investment splurge as classic, trickle-down economics: More tourists mean more money, hence more jobs. The question is at what cost and to whose benefit?

According to Mexico's National Tourism Ministry (Sectur), income from the industry rose from $201.3 million to $487.5 million between 2000 and 2007. Using bigger numbers than Mexico's federal government, the FCCA estimated that cruise ships contributed $565 million and created 16,000 jobs in Mexico during 2006.

The cash flow from cruise ships is but a small slice of a vast international tourism industry that earned Mexico $12.9 billion last year, and employed about 2.4 million people, according to Sectur.

Two lines, Carnival Cruise Line and Royal Caribbean, are the undisputed lords of Mexican waters, according to trade industry and media reports. Although the cruise lines run out of U.S. ports and are traded on the New York Stock Exchange, the foreign-flagged companies are officially based abroad and not subject to most U.S. taxes or labour laws. Indeed, many of the largely unregulated cruise industry's 150,000 workers are from developing nations. They often sleep in hot, squalid underdecks and get $10 to $20 for a 10- to 13-hour work day, according to the International Transport Workers' Federation. Tip earning domestic staff are paid about $50 a month.

Carnival Corporation enjoyed a pair of good years recently. The company's worldwide revenues increased from $11.8 billion in 2006 to $13 billion in 2007, with net income moving up from $2.3 billion to $2.4 billion during the same period. Royal Caribbean's cut of the market grew from $5.2 billion to $6.1 billion, although the world's second largest cruise line watched its net income slip from $633 million to $603 million in the two-year period. Faced with the prospect of slimmer profit margins because of soaring fuel costs in 2008, the industry has begun charging passengers fuel surtaxes up to $140, according to the online publication Cruisecritic.com. Nonetheless, a vibrant European market and a strong euro favour the industry.

But opposition to the expanding industry is also growing. Billionaire and Carnival CEO Micky Arison, who is also the principal owner of the Miami Heat basketball team, was involved with Mexican businessmen in a controversial plan to build a home port on the Mexican Caribbean Island of Cozumel several years ago. The proposed port elicited opposition from environmentalists and hotel owners, and as in Zihuatanejo, the SCT eventually backed down from the project.

Representing 13 lines, the Florida Caribbean Cruise Association participates in meetings and works with Mexican private sector and government leaders to promote the cruise business. The industry recently clashed with Mexican officials over a proposed per-visitor fee for cruise ship passengers. Proponents argued that the charge would help pay for port infrastructure and maintenance; industry countered that added costs would discourage tourism.

Cruise ship companies operate in a "very competitive market" and look for destinations where they can maximize profits, said FCCA President Michele Page in a recent phone interview. That competition means different rules for different ports. Alaska, for example, has slapped a $50 per-visitor fee for environmental and infrastructure services.

Cruise companies already pay their fair share in Mexico, Paige said, given that docking and nearly a dozen other port fees charged can exceed $30,000 per ship. Additionally, cruise companies spend a combined $400 million every year to market Mexico and the Caribbean, Paige said.

After years of wrangling, the Mexican Congress finally decided to levy a $5 visitor fee for cruise ship tourists.

Roberto Maciel, operations manager for the Port of Acapulco, said the fee was originally scheduled to kick in on June 30 but was postponed until September to give Mexican authorities time to implement a collection system. "They are studying the way to collect this," Maciel said.

"It's something that's breathing down our necks. It's not positive, " said Paige. "Costs are extremely high."

Bibliography

Adams, John: *Managing Water Supply and Sanitation in Emergencies.* Oxfam, Oxford, 1999.

Auf der Heide, E.: *Disaster Response: Principles of Preparedness and Coordination,* St. Louis: CV Mosby Co., 1989.

Baird C. : *The Development and Growth of the Cruise Industry,* Oxford, UK: Butterworth-Heinemann, 1999.

Blanton CJ, : *Epidemiology of Gastroenteritis on Cruise Ships Security,* Am J Prev Med., 2006.

Bruce Peter : *Cruise Security: Identity, Design, and Culture,* Rizzoli, 2006.

Cohen, B.: *Deviant Street Networks: Prostitution in New York City,* Lexington, MA: Lexington Books, 1980.

Colbert Judy : *Curise Travel Industry and Safety Management,* Facts on File, 2004.

Cramer EH, Blanton CJ, : *Epidemiology of Gastroenteritis on Cruise Ships,* Am J Prev Med., 2006.

Davey, Kenneth J.: *Elements of Urban Management,* The World Bank, Washington DC, 1993.

Dolcemascolo, Glenn: *Environmental Degradation and Disaster Risk,* Asian Disaster Preparedness Centre, Sweden, 2004.

Donald E. : *Public Personnel Management: Contexts and Strategies,* Upper Saddle River, NJ: Prentice Hall, 1998.

Gibson, Philip : *Crusie Operations Management,* London: Butterworth Heinemann, 2006.

Ines Kwan : *Cruise Control: Regulating Cruise Ship Pollution on the Pacific Coast of Canada,* Vancouver: West Coast Environmental Law, 2001.

Jankowiak, Amy : *Assessment of Cruise Ship Security Effects in Washington,* Olympia, WA : Washington State Dept. of Ecology, 2005.

Kee, Alistair: *Marx and the Failure of Liberation Theology,* Philadelphia, Trinity, 1990.

Kenneth F. : *An Analysis of the Economic Impact of the Cruise Line Industry in Canada,* CCG Consulting Group, 2000.

Kiple, K.F.: *Cambridge World History of Human Disease,* Cambridge University Press, 1993.

Kyle, D. and Koslowski, R.: *Global Human Smuggling: Comparative Perspectives,* Baltimore, Johns Hopkins University Press, 2001.

Marjorie, E.: *Careers in Travel, Tourism, and Hospitality,* Lincolnwood, VGM Career Horizons, 1997.

Mumford, A.: *How Managers can Develop Managers,* Gower, Aldershot, 1993.

Patton, M.Q.: *Qualitative Evaluation and Research Methods,* Newbury Park: Sage Publications, 1990.

Peake DE, Gray CL, : *Descriptive Epidemiology of Injury and Illness among Cruise Ship Passengers,* Ann Emerg Med.. 1999.

Polderman, Jack : *Environmental Protection for Cruise ships - International and Local Requirements,* Lloyd's Register Americas, 2001.

Prasad B V S : *Safety Cruise Tourism : Indian and Global Perspectives,* Icfai University Press, Delhi, 2008.

Sharma Sunil : *Planning and Development of Tourism and Hospitality,* Rajat Pub, Delhi, 2007.

Srikant Kapoor and Prasad B V S : *Cruise Tourism : Indian And Global Perspectives,* Icfai University Press, Delhi, 2008.

Uebersax, Mary B. : *Indecent Proposal: Cruise Ship Pollution in the Caribbean,* 1996.

Vijay Dutt : *Times By The Thames : An Indian Cruise,* Konark Publication, Delhi, 2010.

Wheeler R. : *Travel Health at Sea: Cruise Ship Medicine,* New York: John Wiley and Sons; 2001.

Index

❑❑❑